Poster Sessions 2022

Held at the 2022 AIChE Annual Meeting

Phoenix, Arizona, USA
13-18 November 2022

ISBN: 978-1-7138-7920-6

Printed from e-media with permission by:

Curran Associates, Inc.
57 Morehouse Lane
Red Hook, NY 12571

Some format issues inherent in the e-media version may also appear in this print version.

AIChE
120 Wall Street, FL 23
New York, NY 10005-4020

Phone: (800) 242-4363
Fax: (203) 775-5177

www.aiche.org

Additional copies of this publication are available from:

Curran Associates, Inc.
57 Morehouse Lane
Red Hook, NY 12571 USA
Phone: 845-758-0400
Fax: 845-758-2633
Email: curran@proceedings.com
Web: www.proceedings.com

TABLE OF CONTENTS

On the Linearity of Linear Free-Energy Relationships (LFER) and the Physical Meaning of LFER Parameters

Costas Panayiotou, Ioannis Zuburtikudis, Hadil Abu Khalifeh

For decades, now, the scientific community enjoys the remarkable success of the Abraham solvation parameter model or the linear free energy relationships (LFER) predictive approach. Over the years, LFER has been established as a robust predictive or screening tool for obtaining quantitative structure–property relationships (QSPR) for a broad variety of chemical, biomedical and environmental processes [1-5]. In this model, free-energy-related properties, F, of a solute are correlated to its six molecular descriptors, *V, L, E, S, A*, and *B*, corresponding to the McGowan's characteristic volume, the gas–liquid partition coefficient on hexadecane at 298 K, the excess molar refraction, the dipolarity/polarizability, the hydrogen bond acidity, and hydrogen-bond basicity, respectively. The general correlation is of the linear form: $\log(F) = c + eE + sS + aA + bB + vV + lL$ and quantifies the solute transfer between two condensed phases or from the gas phase. The coefficients (lower case letters) in these equations, which are solvent (phase or system) descriptors, are usually determined by involved multiple linear regression and fitting experimental data. This fitting practice is needed, although the coefficients are attributed (and do have) a widely accepted physical meaning. In fact, no fundamental / thermodynamic explanation is yet available for the very linearity of this established LFER correlation.

In this communication, a clear thermodynamic explanation of this linearity is reported, for the first time. For this purpose, a most simple statistical thermodynamic equation-of-state model [6-11], though which not only the linearity is explained but, also, quantitative equations for the LFER coefficients of the solvent are reported in terms of the solvent-own molecular descriptors. Of particular interest are the equations of a and b in terms of the descriptors A and B, which open new ways for straightforward development and use of LFER-type approaches. Through these equations one may extract information on acid – base or hydrogen-bonding interactions from the freely available databases for thousands of compounds. They also act as tool for identifying eventual outliers in these databases. Various examples of such calculations are reported and the implications for the rational development of unified universal acidity and basicity scales are critically discussed.

Acknowledgements

The financial support by the ASPIRE Award for Research Excellence (AARE-2021) Program of the U.A.E. through grant no. AARE20-246 to Ioannis Zuburtikudis (PI) and Hadil Abu Khalifeh (Co-PI) is greatly acknowledged.

References

1. J. Kamlet, J.L.M. Abboud, R.W. Taft, Proc. Phys. Org. Chem. 13 (1981) 485.
2. Kamlet, M. J.; Doherty, R. M.; Abboud, J.-L. M.; Abraham, M. H.;Taft, R. W.. Chemtech 1986, 16, 566-576.
3. Abraham, M. H., *Soc. Reviews*, **1993**, *22*, 73.
4. Abraham, M. H., Ibrahim, A., Zissimos, A. M., *Chromatogr. A*, **2004**, *1037*, 29.
5. Abraham, M. H., Smith, R. E., Luchtefeld, R., Boorem, A. J., Luo, R., Acree Jr., W. E., *Pharm. Sci.*, **2010**, *99*, 1500.
6. Panayiotou, C.; Sanchez, I. C. Phys. Chem. 1991, 95, 10090−10097

7. Panayiotou, C.. Tsivintzelis, I.. Economou, I., Ind. Eng. Chem. Res. (2007) 46, 2628-2636
8. Panayiotou, C.; Mastrogeorgopoulos, S.; Ataman, M.; Hadadi, N.; Hatzimanikatis, V. Chem.Chem. Phys. 2016, 18, 32570−32592
9. Mastrogeorgopoulos, S.; Hatzimanikatis, V.; Panayiotou, C.. Eng. Chem. Res. 2017, 56 (38),10900−10910.
10. Mensitieri, G. Scherillo, G.. Panayiotou, C.. Musto, P., Materials Science andamp; Engineering R (2020) 140, 100525
11. Panayiotou, C., Hatzimanikatis, V., Fluid Phase Equilibria 527 (2021) 112828-112836

173b Waste Plastic Upcycling: Microkinetic Modelling of Hdpe Pyrolysis

Aswathy Raghu, Rebecca Harmon, SriBala Goruganti, Linda Broadbelt

Plastic pollution is currently a matter of grave environmental concern. Polyethylene (PE) is one of the most widely used plastics due to its applications in packaging and cost-effectiveness. Although mechanical recycling of PE is possible, it becomes infeasible after a few cycles beyond which chemical upcycling to higher value products is a preferred option. Pyrolysis, which is heating of the sample in an inert atmosphere, is a promising way to upcycle PE and thereby produce fuels and chemicals. PE pyrolysis is known to yield gaseous and liquid products, and the product composition strongly depends on reactor operating conditions. Kinetic models developed against experimental data are useful for predicting the reaction characteristics. Global kinetic models available for PE are known to capture overall conversion. Nevertheless, for gaining insights into the reaction mechanism, a microkinetic model is necessary. During pyrolysis, a large number of species of different chain lengths form, and it is computationally very intensive to track every species in a truly mechanistic model. Therefore, it is often more realistic to work with a lumped mechanistic model that tracks 'classes' of species having similar properties, such as dead species, end radicals, mid radicals, etc. Each lumped species class has a molecular weight distribution, and their moments and thereby their properties are calculated by solving population balance equations using method of moments. Physically significant kinetic parameters obtained from theoretical calculations and experiments reported in literature are used in the model. Such mechanistic models for pyrolysis of several polymers have been published by the Broadbelt group. A microkinetic model for HDPE pyrolysis validated using experimental data in literature and its features will be discussed.

173d Multiscale Modeling Approach for Designing Novel Hierarchical Carbon Cathodes for Ultrahigh Capacity Aprotic Li-O_2 Battery

Khizar Hayat, Daniel Bahamon, Lourdes Vega, Ahmed Al Hajaj

In response to mitigate climate change and transition to renewable energy, there is a growing need for energy storage devices to decouple energy growth and transportation sector from CO_2 emissions. Among the myriad electrochemical storage technologies, the Li-O_2 battery is being considered an attractive class of power storage devices because of its high energy density (energy/volume) and specific energy (energy/weight). However, there are few hurdles in its viable commercialization, specifically, lower discharge capacity is one of the main challenge which is induced by the sluggish kinetics and formation of insoluble lithium peroxide (Li_2O_2) inside cathode pore system [1].

Owing to current challenges and limitations, improving the discharge capacity of Li-O_2 battery (LOB) have become critically acclaimed topic among the scientific community. Using experimental studies [2], particular attention has been paid to the optimization and designing of novel cathode materials to overcome the challenges of discharge product deposition. However, multiphysics/multiscale processes occurring in LOB are very difficult to be captured by experimental approaches. Therefore, utilization of a multiscale modeling technique is indispensable to capture various phenomena happening at varying time and length scale, from molecule to system level [3]. The multiscale modeling approach is spanned over the continuum models at macroscopic level to classical forcefield models, reactive forcefield approaches, and the fundamental quantum chemical models (DFT, and ab-initio MD) at the microscopic level. Quantum studies (such as DFT and AIMD) are useful techniques to investigate the reaction-diffusion mechanism, and thermodynamic characteristics during the Li-O_2 battery operation [4]. Unfortunately, such atomistic dynamic studies are restricted to small length scales (~ 100 atoms) and time scales (~ fs to ps). Alternatively, classical forcefield are quite inexpensive computationally and used to model systems at larger time and length scales. However, they do not consider bond breakage or formation during molecular interactions, and hence cannot be employed for reactive systems. Reactive forcefield (reaxFF), firstly developed by van Duin [5], is a bond order based forcefield which has potential to describe the chemical bonds formation and breakage. ReaxFF, like classical forcefields, also incorporates the van der Waals and Coulombic interactions among the molecules. Therefore, ReaxFF are being used to bridge the gap between quantum studies and classical forcefield methods.

In this work, we have developed multiscale top-down approach, in which we started from continuum modeling of Li-O_2 battery cell [6]. We propose different hierarchical cathode designs with distributed tortuosity, and distributed porosity which led to improved discharge capacity by approximately 50 %. It shows that changing pore size and volume, and pore alignment has higher impact on the species (Li, O_2) transport and solid discharge product (Li_2O_2) formation leading to varying LOB performance. Further, we have observed that Li_2O_2 deposition mechanisms (i.e., surface passivation and solution growth) not only depends on the cathode architecture but also the discharge current density. Higher discharge current density leads to film like deposition of Li_2O_2 on the cathode which induces surface passivation. On contrary, lower discharge current density promotes the growth of Li_2O_2 in large toroidal form leading to pore clogging. Considering the continuum modeling findings, it is imperative to dig deep and uncover the impact of structural properties on the electrochemical kinetics, Li_2O_2 deposition and distribution, and species diffusion barriers. For this, we moved one step down towards reactive forcefield based molecular modeling (reaxMD) study of the cathode structural properties and their impact on battery performance.

In our study, we have employed zeolite templated carbons (ZTCs) for LOB cathodes owing to their ultrahigh surface area, and ordered micro-porous framework. In previous work, ZTCs have been employed as secondary battery electrode [7] due to their relatively high conductivity and tunability of structure, however, their potential for LOB cathode never explored. We have performed reactive molecular dynamics simulations of different structures of ZTC such as FAU-ZTC and MFI-ZTC to exploit the diffusion coefficients of the species (including DMSO, $LiPF_6$, O_2, and Li_2O_2), discharge reaction pathways, and deposition of Li_2O_2. The single unit cells were taken from ZTC materials cloud database [8]. Moreover, the effect of hierarchical micro-mesopores was evaluated. The resulted structures (Figure 1, a) lead to enhanced specific surface area as well as pore volume. For the obtained Li_2O_2 system, the basic structural characteristics were studied by pair distribution functions for Li-O and O-O (calculated by reaxFF) and compared with the DFT study [9]. The position of the first peaks in both graphs (Figure 1, b and c) shows the Li-O (i.e., 1.925 Å) and O-O (i.e., 1.275 Å) neighbor distances within the first coordination shell. Compared to Li-O pair, the narrow distribution of O-O pair in first coordination shell in Li_2O_2 structure is because of the presence of unique O-O covalent bond of O_2^{-2}. This narrow pair distribution gives the O-O bond length of Li_2O_2 within the range of 1.275-1.375 Å. On contrary, the broader pair distribution of Li-O is attributed to the coupling of Li^+ with various O_2^{-2}. orientations leading to the Li-O bond distances within the range of 1.775-2.875 Å. Furthermore, the self-diffusivity results of Li-ion and oxygen reveals that, prior to discharge process, both species move randomly through the tunnels of FAU-ZTC which results in quite similar diffusivities i.e., 4.5E-09 for oxygen and 5E-09 for Li-ions along x-direction. However, as discharge process continues and Li_2O_2 forms, the movement of both Li-ion and oxygen become restricted due to the formation of chemical bonds between Li-ion and oxygen. this effect can be further refined by running the simulation for longer time duration.

In conclusion, from our multiscale simulation study, we have deduced that O-O bond length might have impact on the crystalline structure of Li_2O_2, which restricts species transport upon high growth. Therefore, ordered micro-mesoporous FAU-ZTC can provide enough space to store large amount of Li_2O_2 without restricting the flow of ions and hence prolong the discharge process. This highlights the need for multiscale modeling approach that guides the design and selection of materials with enhanced electrochemical properties in terms of energy density.

Fig. 1: Schematics of (a) pore size distribution in FAU-ZTC, and pair distribution of (b) Li-O and (c) O-O inside Li_2O_2. Here DFT studies are taken from [9]

References

[1] Z. Ma, X. Yuan, L. Li, Z.F. Ma, D.P. Wilkinson, L. Zhang, J. Zhang, Energy Environ. Sci. 8 (2015) 2144–2198.

[2] J.W. Jung, S.H. Cho, J.S. Nam, I.D. Kim, Energy Storage Mater. 24 (2020) 512–528.

[3] K. Hayat, L.F. Vega, A. AlHajaj, Renew. Sustain. Energy Rev. 154 (2022) 111849.

[4] X. Luo, R. Amine, K.C. Lau, J. Lu, C. Zhan, L.A. Curtiss, S. Al Hallaj, B.P. Chaplin, K. Amine, Nano Res. 10 (2017) 4327–4336.

[5] M.M. Islam, V.S. Bryantsev, A.C.T. Van Duin, J. Electrochem. Soc. 161 (2014) E3009.

[6] K. Hayat, L.F. Vega, A. AlHajaj, J. Electrochem. Soc. 168 (2021) 120534.

[7] J. Miao, Z. Lang, T. Xue, Y. Li, Y. Li, J. Cheng, H. Zhang, Z. Tang, Adv. Sci. 7 (2020) 2001335.

[8] E.E. Taylor, K. Garman, N.P. Stadie, Chem. Mater. 32 (2020) 2742–2752.

[9] K.C. Lau, D. Qiu, X. Luo, J. Greeley, L.A. Curtiss, J. Lu, K. Amine, Energies. 8 (2015) 529–548.

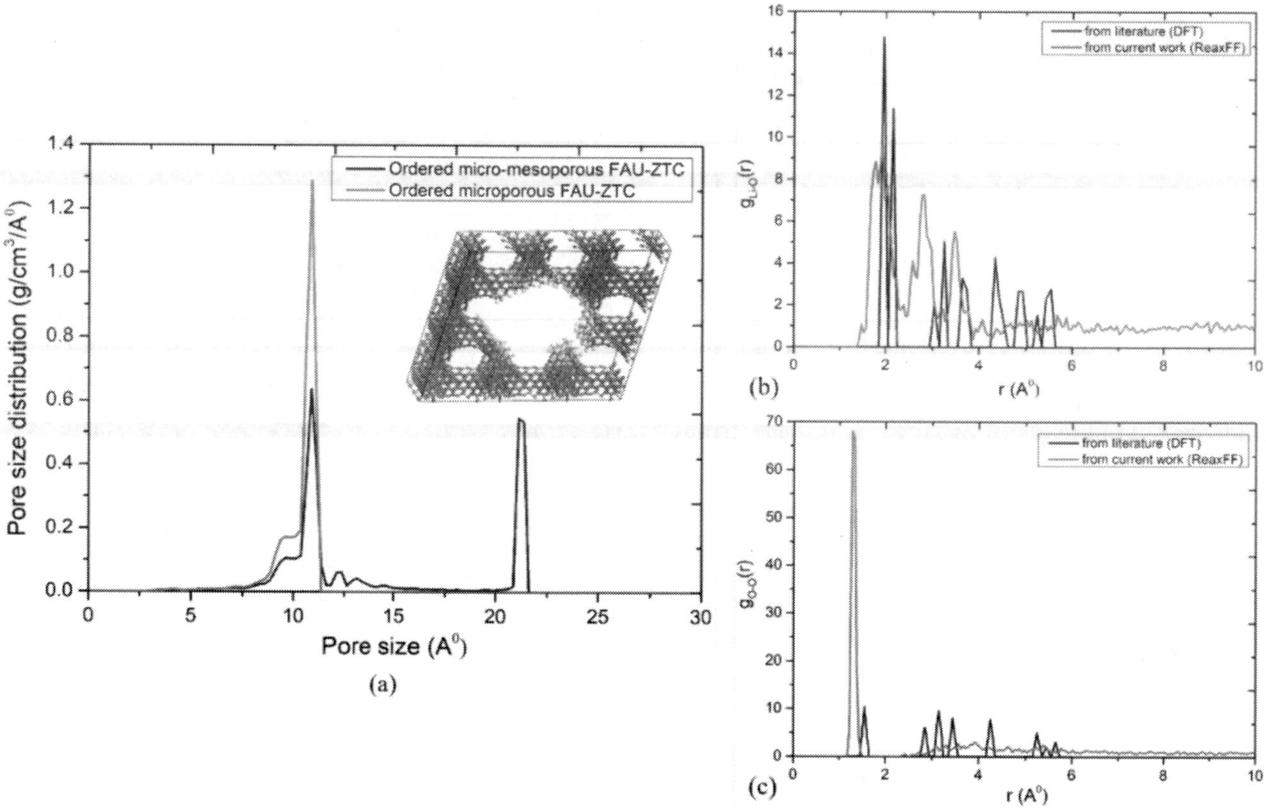
Ordered micro-mesoporous FAU-ZTC
Ordered microporous FAU-ZTC
Pore size distribution (g/cm3/A0)
Pore size (A0)
(a)
from literature (DFT)
from current work (ReaxFF)
gLi-O(r)
r (A0)
(b)
from literature (DFT)
from current work (ReaxFF)
gO-O(r)
r (A0)
(c)

173f Demonstrating the Affect of Polymer Type on Oil Yield in Thermal Depolymerization Using Machine Learning

Elizabeth Belden, Owen Ferrara, Matthew Rando, Eric Himebaugh, Christopher Skangos, Randy Paffenroth, Nikolaos Kazantzis, Michael T. Timko

Chemical recycling of plastics into fuels via processes such as pyrolysis and hydrothermal liquefaction (HTL) is a potentially viable way to handle the millions of tons of waste plastic generated globally. Thermal depolymerization reactions, especially in the presence of mixed plastic feeds, are comprised of complex reaction networks that make traditional physic-based modeling immensely difficult. Data-driven models, with careful model validation and testing, can provide an alternative to the traditional approach. In this study, a dataset consisting of 370 data points of both pyrolysis and HTL of plastic feeds was collected from the open literature. The data was first evaluated for pyrolysis conditions only, resulting in a dataset size of 310 data points. The dataset was divided into training and test sub-datasets and used to evaluate the performance of 7 different machine learning algorithms. Of these models Random Forest Regression and eXtreme Gradient Boosting (XGBoost) predicted the oil yield of the test set with the highest accuracy. They predicted the oil yield for this never-before-seen data with mean absolute error (MAE) values of 8.3% (Random Forest) and 8.7% (XGBoost), respectively. Using the same procedure, the whole dataset (including HTL data) was re-evaluated with the finding that the MAE corresponding to the Random Forest model was 8.5%, similar to that observed for pyrolysis data on its own. The optimized Random Forest model was then used to analyze the oil yield benefit of removing dense (and potentially problematic) plastics, such as polyethylene terephthalate (PET) and polyvinyl chloride (PVC), from pyrolysis feeds. Figure 1 shows the negative impact these polymers have on oil yields in various plastic mixtures.

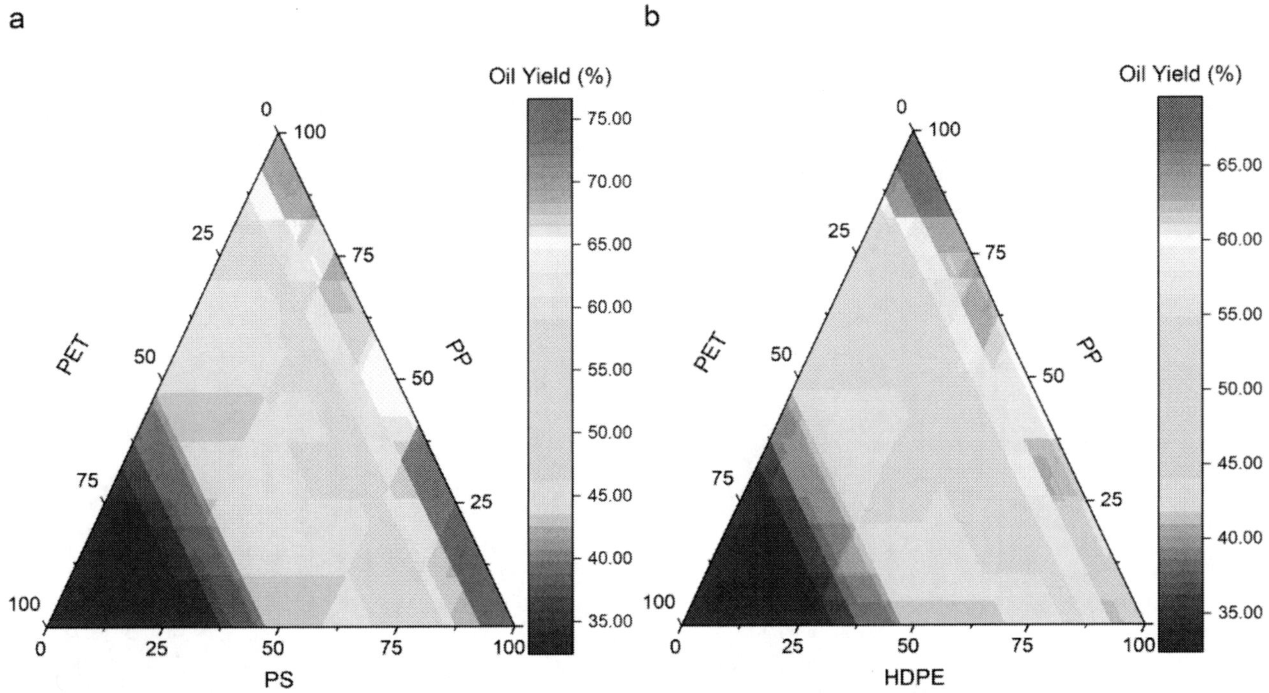

Figure 1 Effect of polyethylene terephthalate on the oil yield of plastic mixtures with a.) high-density polyethylene and polypropylene and b.) polystyrene and polypropylene

173h Tuning C-Phycocyanin Photoactivity Via pH-Mediated Assembly–Disassembly

Ying Li, Alireza Abbaspourrad, Richard Gillilan

Environment-triggered protein conformational changes have garnered wide interests in both fundamental research, for deciphering in vivo acclimatory responses, and practical applications, for designing stimuli-responsive probes. Here we propose a protein–chromophore regulatory mechanism that allows manipulating the photoactivities of C-phycocyanin (C-PC) by environmental pH and UV irradiation.

C-phycocyanin (C-PC), a major protein–chromophore complex originating in cyanobacteria and red algae, is characterized by hierarchical protein architecture and unique light-sensing ability. Using small-angle X-ray scattering, a pH-mediated assembly–disassembly pathway of C-PC in solution state was unraveled. With pH changes in the 3.0–9.0 range, the partially unfolded monomers found in both acidic (pH 3.0) and alkaline (pH 9.0) conditions can be converted to the well-folded trimers at pH 7.0 and high assemblies—a mixture of trimers, hexamers, and nonamers—at pH 5.0. Such flexible protein matrices impart tunability to the embedded tetrapyrroles, whose photochemical behaviors were found to be modulated by protein assembly states. 365-nm UV irradiation on C-PC triggers pH-dependent singlet oxygen ($^1O_2$) generation and conformational changes. Intermolecular photo-crosslinking occurs at pH 5.0 via dityrosine species, which bridges solution-based C-PC oligomers into unprecedented dodecamers and 24-mers. These supramolecular assemblies impart C-PC at pH 5.0 significantly enhanced $^1O_2$ yield, fluorescence emission, and photostability relative to those at other pH values.

The identified pH tunability via protein–chromophore ineractions makes C-PC a potential pH-responsive photosensizer for tumor-targeted photodynamic therapy. The enhanced activation of C-PC in weakly acidic environments (pH 5.0) makes C-PC a potential pH-responsive photosensizer for tumor-targeted photodynamic therapy as tumor microenvironments, for example, the lysosome (pH ~4.5–5.0), are characterized by similar acidity. In this sense, C-PC could be utilized in the design of novel pH-responsive fluorescent imaging probes or $^1O_2$ generators to boost photodynamic activities at acidic tumor sites while minimizing phototoxicity to non-target healthy tissues. It is envisaged that with further development, the remarkable pH-responsiveness, photoactivity, and conformational flexibility of C-PC demonstrated in this work, along with this protein's inherent biocompatibility and wide availability, will open new routes for its applications in cancer diagnosis, imaging, and photodynamic therapy (PDT).

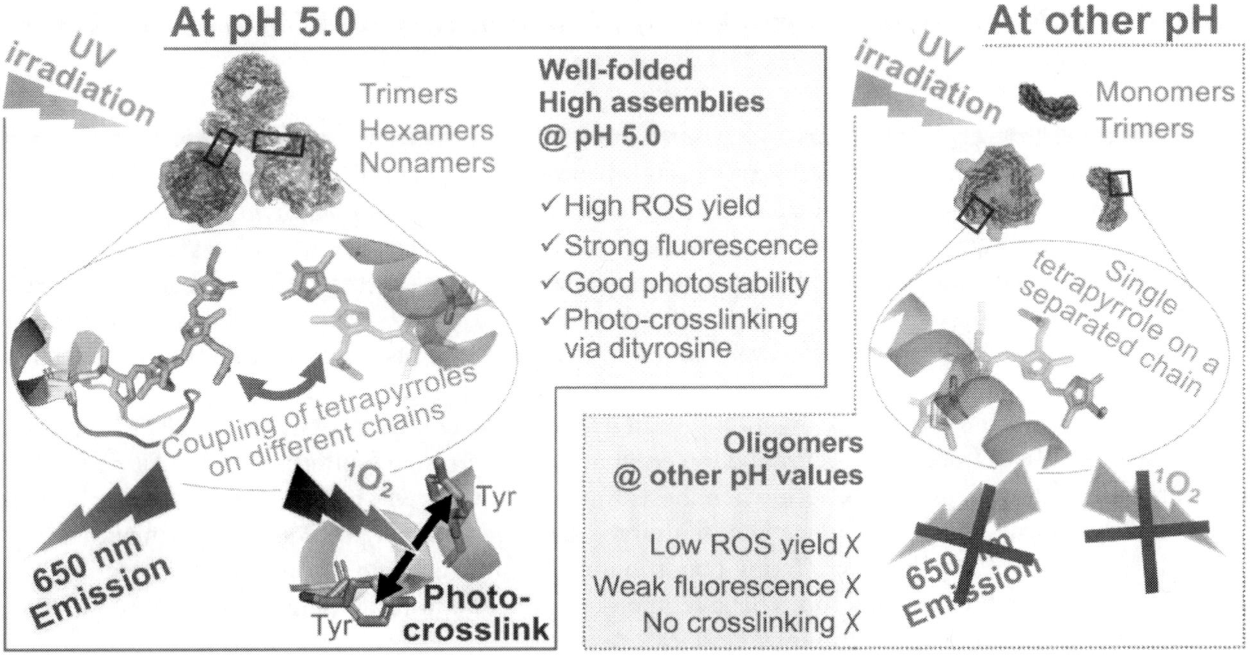
At pH 5.0
UV irradiation
Trimers
Hexamers
Nonamers
Well-folded
High assemblies
@ pH 5.0
✓High ROS yield
✓Strong fluorescence
✓Good photostability
✓Photo-crosslinking via dityrosine
Coupling of tetrapyrroles on different chains
650 nm Emission
$^1O_2$
Tyr
Tyr
Photo-crosslink
At other pH
UV irradiation
Monomers
Trimers
Single tetrapyrrole on a separated chain
Oligomers
@ other pH values
Low ROS yield ✗
Weak fluorescence ✗
No crosslinking ✗
650 nm Emission
$^1O_2$

173i Study on Growth and Metabolite Trend Comparison of *Staphylococcus Aureus* and *Pseudomonas Aeruginosa* in Monoculture and Co-Culture System

Heejoon Park

One of the most prevalent types of multi-microbial infection occurs in chronic wounds associated with diabetes, cystic fibrosis, and venous disease. *Pseudomonas aeruginosa* and *Staphylococcus aureus* are two of the most common bacteria isolated from chronic wounds and are frequently found together, which implies they possess synergistic interactions in the wound environment. Multi-species infections are more virulent and often result in worse patient outcomes than single microbe infections. The microbes often express a biofilm phenotype which has strong tolerance to numerous interventions such as antibiotics. The goal of this project is to understand microbes interaction, and mechanisms in planktonic culture during monoculture and co-culture condition. Our initial investigation revealed that the co-culture system showed 277% increase in growth rate and 138% more biomass production comparing to their monoculture. This indicate that co-culturing system showed growth benefit when these are coexist in growth media. Also their metabolic trend reveals that syntrophy effect takes place in co-culture system by different preference on primary carbon source. It led more stable in lower buffering condition in co-culture system. From this investigation, we will have better understanding how microbes behave in co-culture conditions and the results will be beneficial to find the better approach to treat chronic wound infections.

173j A Deep Learning-Based Feature Extraction Framework for Monitoring High-Order Nonstationary Industrial Processes

Cheng Ji, Tingting Tao, Fangyuan Ma, Jingde Wang, Wei Sun

Early detection of abnormal process deviations is an urgent need for modern manufacturing plants. When distributed control system is widely used to reach the effective control of process, a large number of process variables can be measured and stored, which greatly facilitates the development of data-driven process monitoring technologies. Numerous multivariate statistical and artificial intelligence methods have been employed to establish process monitoring models by extracting internal correlations among variables. However, most reported methods are developed by assuming that the process is stationary, while industrial operation data show high-order nonstationary characteristics, as the statistical indicators, such as the mean and standard deviation of certain variables, change with time. Differencing is a common and effective way of dealing with nonstationary sequences in the field of economics, but the data processed after more than second-order differencing will lack interpretability. Moreover, useful process information, especially the small changes at the early stage of faults, can be removed simultaneously by differencing, making the fault difficult to be early detected. Cointegration theory is an alternative to nonstationary modeling by extracting the long-term equilibrium relationship among nonstationary variables. However, it is still not applicable to industrial processes because it is still limited to the extraction of linear relationships.

In this work, a novel process monitoring framework is proposed by integrating the idea of differencing with a deep encoder-decoder recurrent neural network. Considering that the forget gate of most memory neural networks is saturated with long-term stationary information, and the short-term nonstationary characteristics cannot be effectively extracted, nonstationary information obtained by differential transformation is input to the next memory neural network structure in this work, and therefore the forget gate can be used to choose how much nonstationary information to forget or remember. A stacking multilayer network is constructed to effectively extract high-order nonstationary characteristics in industrial processes. The process nonlinearity and dynamic can be extracted as well by utilizing recurrent neural network. The encoder-decoder structure is employed to perform unsupervised fault detection. The mean square error and KL divergence are integrated as the monitoring index of the proposed method. Case studies on both the Tennessee Eastman process and an industrial continuous catalytic reforming heat exchange unit are presented to illustrate the effectiveness of the proposed method. The results confirm that the proposed method can be applied to early detect the faults with a low false alarm rate by effectively extracting the high-order nonstationary characteristics of industrial processes.

173k Machine Learning Analysis of Multimodal Data from a Smartphone-Based Electrochemiluminescence Sensor.

Hyun Kwon, Elmer Ccopa Rivera, Rodney Summerscales

Machine learning was applied for the data analysis of multimodal data from a smartphone based electrochemiluminescence (ECL) sensor. This portable sensor with low-cost alternative technology was used for the detection and quantification of phenolic compounds. The effective quenching of the ECL reaction by the phenolic compounds, vanillic acid and p-coumaric acid, was recorded by the signals of light intensity and electric current using apps of the sensor. Due to common problems present in sensor data such as non-linearity, multimodality, sensor-to-sensor variations, presence of anomalies, and ambiguity in key features, several machine learning strategies were explored. In contrast to the traditional calibration approach of extracting predetermined key features, the ML methods such as tri-layer neural net or boosted trees carried out effective regression tasks by learning higher patterns without pre-processing the key features. Combined multimodal features made 80% enhanced performance with multilayer neural net algorithms than the traditional approaches. The results demonstrated that the ML methods could provide a robust analysis framework for sensor data with noises and variability without preprocessing to extract features or examine ambiguous anomalies.

173o *Production of Novel Sars-Cov-2 Spike Truncations in Chinese Hamster Ovary Cells Leads to High Expression and Binding to Antibodies*

Shiaki Minami, Seongwon Jung, Yihan Huang, Bradley Harris, Matthew Kenaston, Roland Faller, Somen Nandi, Karen A. McDonald, Priya Shah

Severe acute respiratory syndrome coronavirus 2 (SARS-CoV-2), which causes coronavirus infectious disease 2019 (COVID-19), has led to over 6 million deaths to date. Containment efforts such as surveillance and herd immunity require generating large quantities of viral glycoproteins. Therefore, the COVID-19 pandemic has highlighted the critical need for rapid, scalable, and cost-effective production of recombinant glycoproteins for use as antigens in clinical and research applications.

For SARS-CoV-2, scalable production of Spike glycoprotein is important for diagnosis and vaccination strategies. Spike is a major antigen that is responsible for entry into cells, and it is the primary target for antibody binding. Thus, immunoassays to assess immunity of individuals commonly utilize the Spike protein. Protein-based SARS-CoV-2 vaccines also rely on delivering Spike protein (Heath et al., 2021).

A major challenge to scaling these immunoassays and protein-based vaccines is producing large amounts of Spike protein in a cost-effective manner. Different forms of full-length Spike have been produced in mammalian cell lines, including mutations to increase stability, but Spike is difficult to express, with titers typically at approximately 5-30 mg/L (Amanat et al., 2020; Hsieh et al., 2020). Alternatively, the Spike RBD has been expressed, which can have titers that are an order of magnitude above those of Spike. However, RBD is less sensitive than Spike in serological assays (Amanat et al., 2020). Modifications to Spike and RBD have been made to increase stability and expression, such as through rational structure-based approaches and mutational scanning (Hsieh et al., 2020; Smaoui andamp; Yahyaoui, 2021; Starr et al., 2020). Nonetheless, conserving the sequence of Spike and RBD is important for accurate comparisons with existing variants.

Here, we improve transient Spike expression in Chinese hamster ovary (CHO) cells. We optimize the production period of Spike and demonstrate that Spike titers increase significantly over the expression period, maximizing at 14 mg/L at day 7. In comparison, RBD titers peak at 54 mg/L at day 3. We also develop 8 Spike truncations (T1-T8) to determine if a truncation can balance the high expression of RBD and high binding capabilities of Spike. The truncations T1 and T4 have high expression at 130 mg/L and 73 mg/L, respectively, which are higher than RBD titers. We also evaluate antibody binding via ELISA against antibodies raised against full-length Spike. T1 has similar sensitivity as Spike against an anti-Spike monoclonal antibody and even outperforms Spike for an anti-Spike polyclonal antibody. T4 has overall lower sensitivity than full-length Spike to the antibodies, but still has higher sensitivity than RBD.

Using molecular dynamics simulations, we find that T1 has more structural similarity to full-length Spike than does T4. The computational results are corroborated by circular dichroism analysis for secondary structure composition, which also show higher similarity between Spike and T1 than T4. Taken together, these results suggest T1 is a promising Spike alternative for use in various applications. Future investigations include serological tests using the truncations.

References

Amanat, F., Stadlbauer, D., Strohmeier, S., Nguyen, T. H. O., Chromikova, V., McMahon, M., Jiang, K., Arunkumar, G. A., Jurczyszak, D., Polanco, J., Bermudez-Gonzalez, M., Kleiner, G., Aydillo, T., Miorin, L., Fierer, D. S., Lugo, L. A., Kojic, E. M., Stoever, J., Liu, S. T. H., ... Krammer, F. (2020). A serological assay to detect SARS-CoV-2 seroconversion in humans. *Nature Medicine*, *26*(7), 1033–1036. https://doi.org/10.1038/s41591-020-0913-5

Heath, P. T., Galiza, E. P., Baxter, D. N., Boffito, M., Browne, D., Burns, F., Chadwick, D. R., Clark, R., Cosgrove, C., Galloway, J., Goodman, A. L., Heer, A., Higham, A., Iyengar, S., Jamal, A., Jeanes, C., Kalra, P. A., Kyriakidou, C., McAuley, D. F., ... Toback, S. (2021). Safety and Efficacy of NVX-CoV2373 Covid-19 Vaccine. *New England Journal of Medicine*, *0*(0), null. https://doi.org/10.1056/NEJMoa2107659

Hsieh, C.-L., Goldsmith, J. A., Schaub, J. M., DiVenere, A. M., Kuo, H.-C., Javanmardi, K., Le, K. C., Wrapp, D., Lee, A. G., Liu, Y., Chou, C.-W., Byrne, P. O., Hjorth, C. K., Johnson, N. V., Ludes-Meyers, J., Nguyen, A. W., Park, J., Wang, N., Amengor, D., ... McLellan, J. S. (2020). Structure-based design of prefusion-stabilized SARS-CoV-2 spikes. *Science (New York, N.y.)*, eabd0826. https://doi.org/10.1126/science.abd0826

Smaoui, M. R., andamp; Yahyaoui, H. (2021). Unraveling the stability landscape of mutations in the SARS-CoV-2 receptor-binding domain. *Scientific Reports*, *11*(1), 9166. https://doi.org/10.1038/s41598-021-88696-5

Starr, T. N., Greaney, A. J., Hilton, S. K., Ellis, D., Crawford, K. H. D., Dingens, A. S., Navarro, M. J., Bowen, J. E., Tortorici, M. A., Walls, A. C., King, N. P., Veesler, D., andamp; Bloom, J. D. (2020). Deep Mutational Scanning of SARS-CoV-2 Receptor Binding Domain Reveals Constraints on Folding and ACE2 Binding. *Cell*, *182*(5), 1295-1310.e20. https://doi.org/10.1016/j.cell.2020.08.012

Modified Leclanche Cell for Cheap Power Generation

David Judbarovski

Modified Leclanché cell's power system

Modified Leclanché cell's power system

Author: David Judbarovski, judbarovski@gmail.com

Linkedin

My present conceptual invention is a new look on power generation. Here just below I disclose such systems. It generates electricity by USD 0.01/kWh (CAPEX + OPEX), with CO2 exhaust of 0.23 kg/kWh vs. USD 0.09 and 1.0 kg CO2 for coal TPP, or USD 0.09 and 0.46 kg for NG TPP correspondingly, if USD 80.0/ton coal and if domestic USD 100.0/1000 m3 NG. Energy efficiency of electrochemistry was supposed as 70% and for TPP as 40%.

The Leclanché cell is a battery invented and patented by the French engineer Georges Leclanché in 1866. I slightly modified his cell to turn the big problem of his ECG for the earliest age of electrochemistry into pretty advantage for our present level of technologies. Mr. Georges had developed a special technological trick being a usage of composite electrode containing manganese oxide for his Zn-battery to avoid its hydrogen being a byproduct of his battery. In my turn, I offer to utilize such hydrogen for additional electricity production up to 3.7 kWh per 1.0 kg of the Zink vs. about 2.0 kWh/kg Zn if the Leclanché original (theoretically). Moreover, my buttery is cheaper and simpler and much simpler recyclable. Moreover, if being recycled by CH4 instead of the coal, we are adding additionally 2.3 kWh/kg Zn as high temperature water vapor can be converted in additional electricity and sufficiently diminish the CAPEX vs. Zn recycling by coal

Being summarized, my battery
anode reaction: Zn – 2 e (-) = Zn(+2)
electrolyte: 2 NH4(+) + 2 OH(-) + 2 H2O + Zn(+2) = 2 H3O(+) + Zn(OH)2 solid + 2 NH3..
Cathode reaction: 2 H3O(+) + 2 e(-) = H2 + 2 H2O
And totally:
Zn + 2NH4OH = Zn(OH)2 + 2 NH3 + H2 + 2.45 kWh/kg Zn (theoretically)
H2 + air = (Fuel cell) = additional and practically free electricity of 1.22 kWh/kg Zn, so totally 3.7 kWh/kg Zn., while the said H2-air Fuel cell is in order of magnitude cheaper than H2O-electrolizer per kW, so it allows to produce additional electricity practically free.

For comparison, the Zn-air battery is 1.5 kWh/kg Zn (theoretically).

Further we'll thermo-chemically recycle the Zn from Zn(OH)2 consuming the coal or NG both as a chemical and as a energy source, and restore NH4OH from NH3 by it dissolving in the water.

(A) Zn(OH)2 + 0.91* C + n * O2 = Zn + 0.91 CO2 + H2O + 0.0 kJ

Zn(OH)2 + ½ C = 950 C = Zn + ½ CO2 – 163 kJ

and in sum it is 163 kJ, can be added by

$C + O_2 = CO_2 + 393$ kJ, and 163 / 393 = 0.41 C

And totally

$Zn(OH)_2 + (0.5 + 0.41) C + 0.41 O_2 = Zn + (0.5 + 0.41) CO_2$, or

$Zn(OH)_2 + 0.91 C = Zn$ recycled $+ 0.91 CO_2$

If supposing C to be USD 80.0/ton = 8.0 USA cents per kg, so

Zn can be recycled by cost of 8 * 12 * 0.91 /(65 * 3.7 * 0.7) = 0.5 cent USA/kWh + CAPEX

(B) $2 NH_3 + 2 H_2O = 2 NH_4OH - 504$ kJ = 2.15 kWh heat/kg Zn.

The said heat can be created by solar energy at low concentration, e.g. by two mirrors system made of cheap thin foils and providing a motionless focal spot. It can be as cheap as less than USD 20.0/kW, or less than USD 0.0005/kg Zn.

173p Microed: Cryo-Electron Diffraction of 3D Microcrystals

Alison Haymaker, Brent L. Nanenga

One of the critical bottlenecks in crystallography is the growth of large, well-ordered crystals for conventional X-ray experiments. Microcrystal electron-diffraction, or MicroED, is a structure determination method that uses a cryo-transmission electron microscope (cryo-TEM) to collect electron diffraction data from microcrystals. This technique has been successfully used to determine structures of targets from proteins, small organic molecules, and materials, all from crystals orders of magnitude smaller than what is needed for X-ray diffraction experiments. Here, we will describe the MicroED method and its applications, as well as continued improvements to facilitate MicroED structure determination for more difficult crystalline samples.

173q Understanding Protein Mediated Biomineralization Using Cryogenic Electron Microscopy

Sagnik Sen

Biomineralization refers to the process by which living organisms facilitate the formation of minerals, and proteins have been found to be the main mediating agents responsible for biomineralization (1). Elucidating the underlying molecular interactions behind biomineralization is important for fully understanding the fundamental mechanisms responsible, and ultimately harnessing these mechanisms for the design of hybrid biological-nanomaterial systems. Presently, there is a dearth of high resolution structural models of protein- inorganic NP (nanoparticle) interactions (2). Cryogenic electron microscopy or cryo-EM is a structural biology technique that has been used to obtain very high resolution structures of proteins including visualization of individual atoms (3). Cryo-EM can image proteins in their native state and functional environment. In this study we demonstrate the ability of cryo-EM as a promising powerful tool to study the protein - NP interface in biomineralization. We analyzed two systems ; GroEL -Platinum NP complex and Ferritin - Iron NP complex.

REFERENCES:

1. Wang, L. andamp; Nilsen-Hamilton, M (2013), "Biomineralization proteins: From vertebrates to bacteria", Frontiers in Biology 8(2), 234–246.
2. Weidner, T.; Castner, D. G. (2013), "SFG analysis of surface bound proteins: a route towards structure determination", Phys Chem Chem Phys , 15 (30), 12516-24.
3. Yip, K.M., Fischer, N., Paknia, E. et al (2020), "Atomic-resolution protein structure determination by cryo-EM", Nature 587, 157–161.

173s Photoelectrochemical Reduction of Oxygen on Two-Dimensional Covalent Organic Frameworks

Syed Ibrahim Gnani Peer Mohamed, Amir Shariffar, Christopher Merkel, Mona Bavarian, Siamak Nejati

Porphyrinoids are promising materials that have become the subject of much interest in the field of photocatalytic research. In nature, prophyrinoids are the basis for energy generation. They catalyze photosynthesis in plants and cellular respiration in animals. All porphyrinoids contain a tetrapyrrole porphyrin structure, which mirrors the photosynthetic reaction center of chlorophyll. Evolution has deemed this molecular structure the most efficient in converting solar energy into chemical potential energy. Among the porphyrinoids family, porphyrin-based materials are emerging due to their ease of synthetic pathway and functionalization. Polymerization of porphyrins to create covalent organic frameworks (COFs) and networks increases the stability of these materials. Herein, we utilize a vapor phase pathway to deposit thin-films of polymeric materials containing porphyrin-based COFs. The as-synthesized material was employed as a photoelectrocatalyst for oxygen reduction reactions and illuminated under visible light while placed in a basic solution saturated with oxygen. The result was a 50% improvement in cathodic current under illumination, and a shift in the onset potential of reduction, suggesting a shift in the flat band potential of the COFs as a result of the accumulation of photogenerated carriers in the polymeric network. The balance between the ion transfer and the charge transfer enables us to record the dynamics of photo-to-fuel conversion in ORR. The materials were further processed and the coating was utilized as a precursor for creating conductive electrodes hosting single-atom catalysts. We report on the enhanced measured photocatalytic performance and the number of active sites generated via post-processing.

173ah Discovering Latent Effective Parameters from Heterogeneous Populations

Nikolaos Evangelou, George Kevrekidis, Felix Dietrich, Ioannis G. Kevrekidis

We introduce a data-driven framework that discovers and disentangles latent parameters [1] given trajectories of individuals in heterogeneous populations described by parameter dependent ordinary and stochastic differential equations. In our framework we assume that we have access to trajectories from a black box model that does not provide any information of the effective parameters that describe the dynamics of the individual in the heterogeneous population. Our framework involves (a) using the manifold learning technique Diffusion Maps [2] to discover the effective inputs that characterize the individual's behavior and (b) using Conformal Autoencoders Neural Networks [1] as well as the Jointly Smooth Function[3] algorithm to disentangle the latent heterogeneity parameters from the remaining degrees of freedom common. We illustrate that the identified latent parameters are one-to-one with the true heterogeneity of the population parameter. We consider synthetic data generated from parameter dependent ordinary and stochastic differential equations as a proof of concept. We also illustrate the ability of our scheme to identify the latent parameter given experimental trajectories of cellular data with applications to aging.

[1] Evangelou, N., Wichrowski, N. J., Kevrekidis, G. A., Dietrich, F., Kooshkbaghi, M., McFann, S., andamp; Kevrekidis, I. G. (2021). On the Parameter Combinations That Matter and on Those That do Not. *arXiv preprint arXiv:2110.06717.*

[2] Coifman, R. R., andamp; Lafon, S. (2006). Diffusion maps. *Applied and computational harmonic analysis*, 21(1), 5-30.

[3] Dietrich, F., Yair, O., Mulayoff, R., Talmon, R., andamp; Kevrekidis, I. G. (2022). Spectral discovery of jointly smooth features for multimodal data. *SIAM Journal on Mathematics of Data Science*, *4*(1), 410-430.

Potassium Salt Technology Supposing to Reject Fuels Import

David Judbarovski

Potassium Salt Technology supposing to reject fuels import

Author: David Judbarovski, systems engineering, principle inventor

judbarovski@gmail.com

Publicity resume

Technology offered, and well scalable, it comprises two steps, are cycled supposedly 100 cycles for each material unit, here being KNO3 : (1) electrochemistry and (2) thermal decomposition for restoring of first step, the both are very energy effective inherently.

Therefore, they transform inherently cheaper thermal energy in electricity practically free of further cost burdens, and with hundredfold less material consume.

All those create cheap andamp; clean andamp; sustainable electricity.

For example,

To reject 100 milliards cube meters of natural gas imported annually by Germany and can create ~ 65 % annual electricity consume (~400 milliards kWh), our technology would be much cheaper 2.0 cents/kWh (OPEX + CAPEX = 5 years payback), using 5300 km2 (1.5 % of German territory for solar power plants, if 7.0 m2 land area per 1.0 m2 solar flux and a seasonal of 200 days in a year for the said KNO3 thermal decomposition.

Each cube km of oceans contains 0.4 million ton of potassium ions, can be extracted industrially. The more material annual production the less its cost, and its cost slow depends on a kind of material either rare it or common (1980).

DISCLOSURE FOR EXPERTS

(1) KNO3 = (thermolysis) = KNO2 (recycled) + ½ O2 + 124 kJ * 100 cycles / 102 gram/mol = 120 kJ/gram KNO3

(2) KNO2 (water solution) + ½ O2 / air = (flow Fuel cell) = KNO3 (sediment) + 124 kJ or 84 kJ electricity / gram KNO3, if 70% efficiency, so 4.0 times more powerful, than Natural Gas in TPP. Really, ~ 60 kJ/gram * 35 % efficiency = 21 kJ/g + a lot of CO2 exhaust

KNO2 solubility is tenfold bigger than for KNO3, and

KNO2 + H2O = K (+) + (OH) (-) + HNO2, exothermal oxidized in HN03 and then gives KNO3 sediment, extracted periodically for thermolysis in KNO2

Heat of the said thermolysis is created by solar concentrators comprise pairs of small mirrors about 1m2 each and made of thin foils and sheets , so cheap and strong, about USD 20.0 kW with high temperature heat of 0.35 cent/kWh for moderate climate as in Germany. Planar mirror redirects solar flux on dish mirror, so focal spot is motionless.

Really, my alternative technology for analogous solar concentrators with rectenna thermal heterodyne in focal spot for high effective direct transforming such heat in electricity about 0.6 cent/kWh, can be more preferable

173ai Insight to Crystallisation Fouling from 4D X-Ray: A Three Step Process?

Isaac Appelquist, Peter Winkel Rasmussen, Henning Osholm, Anders Nymark Christensen, Anders Bjorholm Dahl, Philip Loldrup Fosbøl, Benaiah Anabaraonye

Crystallisation fouling presents a costly challenge, as it lowers the efficiency of heat exchangers, decreases the throughput of desalination membranes and clogs underground pipelines. Optimal maintenance requires accurate knowledge of the kinetics of the formation of crystallisation fouling.

Current methods for comprehensive in situ investigations are limited. The most commonly used ex situ or 2D experiments cannot capture dynamic effects or describe formation kinetics. This work uses X-ray computed tomography (CT) scanning to address this challenge.

We recorded the formation of crystalline fouling in situ by installing a custom-built flow loop inside a CT cabinet. To record the temporal changes in the crystalline fouling, we had to sparsely sample and underexpose the specimen, so we employed a custom reconstruction algorithm ensured accurate reconstructions. The high spatiotemporal resolution of the experiment made it possible to study the processes of crystal formation.

Our findings show how the growth rate can vary 300 % in the initial phases of growth. Even though we performed measurements for 150 hours, the rate at which growth formed was not constant. This non-equilibrium effect reflects the complexity of crystallisation formation in natural systems. Our findings provides a foundation for unveiling the dynamics of fouling mechanisms, which will aid in developing more accurate prediction models.

173aj Inter-Electronic and Inter-Valley Transitions in MoS_2-WS_2 Heterostructures and Alloys

Sungjoon Kim, Vikas Berry

Van der Waals heterostructures allow the integration of functionalities from different nanomaterials to obtain advanced, superimposed properties. Here, vertical MoS_2-WS_2 heterostructure grains were grown using atmospheric pressure chemical vapor deposition, and their optoelectronic properties were investigated. The field effect transistors fabricated from the heterostructure grains exhibited high photoresponsivity, which was influenced by the gating potential, as well as the measurement temperature. During the gating potential sweep from -5 V to +55 V, the light to dark source drain current ratio (I_{light}/I_{dark}) increased from 3 to 16 and from 2 to 7 at 60 K and 120 K, respectively. Alloy of $Mo_xW_{1-x}S_2$ grains were also obtained, which displayed valley coherence when irradiated with linearly polarized light. The degree of valley coherence was approximately 49% at room temperature. Such phenomena can be leveraged to realize novel optoelectronics, computing, and memory applications.

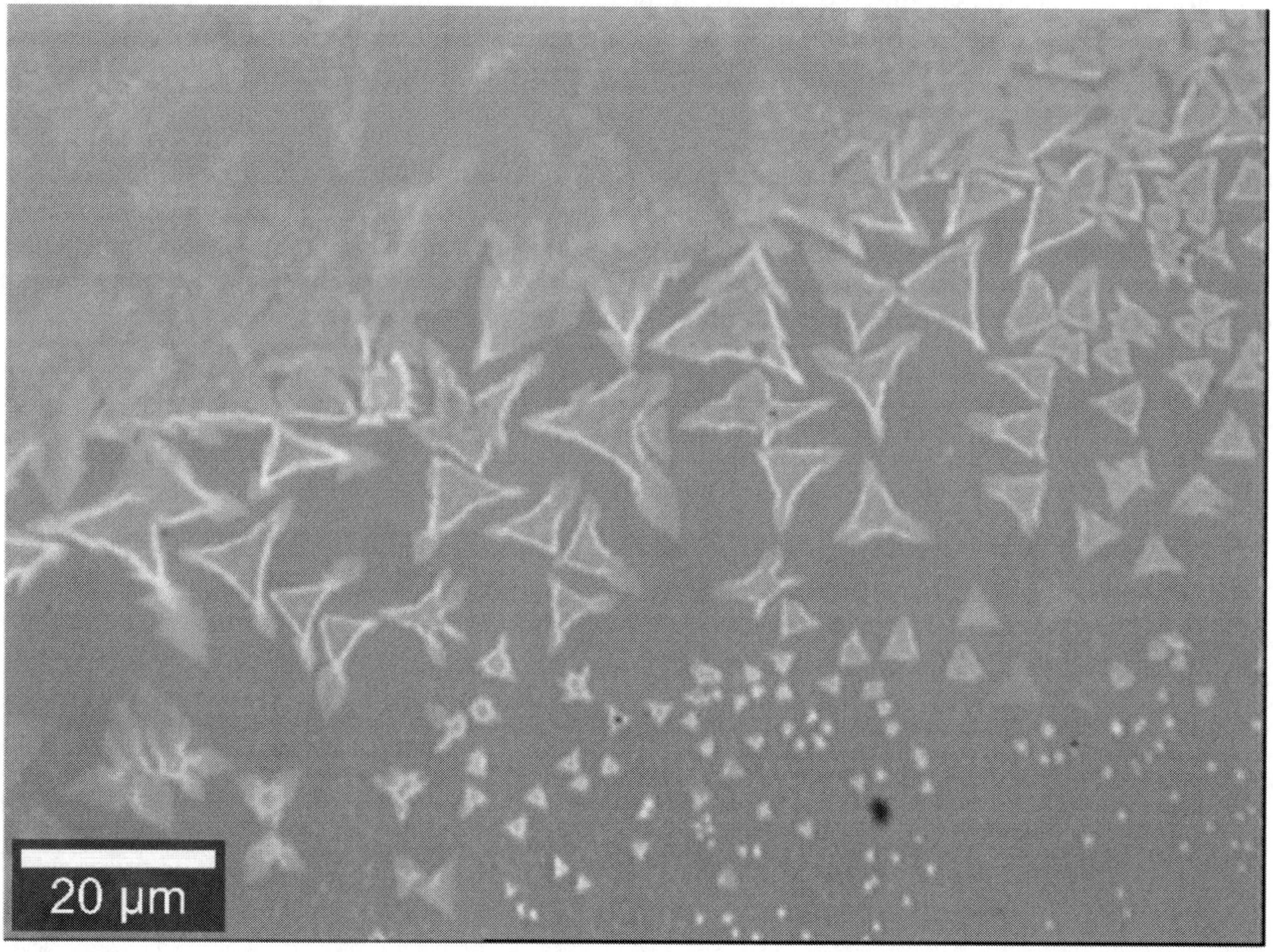

173ak Bacterial Conversion of Methane to Lipids: A Technical Assessment and Benefit-to-Cost Analysis As Methane Abatement Strategy

Lisa Stephanie Dizon, Robert Bertrand, William Holmes, Rafael Hernandez, Mark Zappi, Dhan Lord Fortela, Emmanuel Revellame

The reduction in greenhouse gas emissions (GHG) has become the collective goal of various countries to address the rising global climate change concerns. The United States Environmental Protection Agency proposed a new Clean Air Act rule that would lead to significant, cost-effective reductions in methane emissions in support to the Global Methane Pledge for climate change mitigation. Methane is the major component of natural gas. Natural gas is flared into the atmosphere because of its historic low and volatile prices in the past decade, which represents product and energy losses as well as major GHG emissions. One potential solution is to utilize methanotrophs due to their ability to consume methane as their sole carbon and energy source to produce a wide variety of bio-products like lipids. These bacteria are naturally present in a variety of environment such as wetlands, paddies, sediments, landfills, and waste treatment facilities. The lipids derived from methanotrophs are vital precursors for the production of green fuels and as oleochemical feedstocks. Thus, this research aimed to establish the growth conditions of methanotrophic consortium present in wastewater sludge that optimize methane consumption and lipid production. Growth parameters were varied to study the effects of media type, initial pH and methane to air volume ratio on the bacterial conversion of methane to lipids. Moreover, this study intended to present a science-based evaluation of the benefits and costs of methane utilization for lipid production via social cost benefit analysis. The results highlight the potential of natural microbial consortium to convert methane into lipids using wastewater as medium that could address concerns on GHG emissions as well as for value-added resources recovery. Additionally, these findings could serve as basis for policy recommendations on the potential of methanotrophs as a methane abatement strategy toward climate change mitigation.

Prediction of Mixing Time By Rotational Speed of Bioreactor

Hyunwoo Lee, Youngjin Seo, Joon-Chul Lee

Bioreactor is a device that produces a desired product by growing cells under an appropriate environment after introducing a medium. Methods of culturing cells using the bioreactor is used in various industries, and related research is in progress. In this study, the bioreactor was operated with orbital rotation, and the rotational speed was controlled. For reliability of numerical analysis, results of experiment and numerical analysis were compared. Mixing time was predicted by varying the rotation speed and scale using the verified numerical analysis model.

173al Computational Design of Therapeutic Drug Formulations to Control Protein Interactions

Gregory Dignon

Biologics are very common as theraepeutic drugs, particularly monoclonal antibodies. One downside of biologics is their tendency to phase separate or aggregate at concentrations needed for administration. One strategy to improve the behavior of the drug in solution is to incorporate small molecule additives, excipients, that disrupt interactions between the drug molecules. To find the ideal excipient for a given biologic drug is a very important, yet elusive problem. Here we present a framework to understanding the competition between drug-drug and drug-excipient interactions at the molecular level. This dissects the problem into two problems that are much more manageable for current computational methods.

173am A Novel Application of the Newcomb-Benford Law to Exposure Data

Byron Fuentes, Öyku Dinçkol, Spyros Karakitsios, Dimosthenis Sarigiannis, Silvia Valentini, Elena De Felip, Laura Ricceri, Gemma Calamandrei, Anna Pino, Dayna Schultz

The Newcomb-Benford Law (NBL) is a mathematical concept that describes the distribution frequency of the significant digits for stochastic processes (Li, Fu, and Yuan 2015). It was first noted by Simon Newcomb in 1881 (Newcomb 1881) and later revived by Frank Albert Benford in 1938 (Benford 1938) who applied the concept to a large sample of data sets. The NBL suggests that the probability for the significant digits (1 through 9) of a randomly occurring number is not equally distributed (Equation 1), as one would consider by chance ie.if every digit were to demonstrate equal probability, each digit should have a relative frequency of 11.1% (ie each of the 9 digits have equivalent probability). However, the NBL describes a distribution where 1 has a higher relative frequency than 2 and so on.

The use of the NBL has found widespread applications across many fields even though the underlying reasons for its prevalent applicability still remain somewhat of a mystery (Li, Fu, and Yuan 2015). Particularly, the NBL has been widely used to identify fraudulent or manipulated data in the political (voter fraud; Pericchi and Torres 2011), economic (tax evasion; Demir and Javorcik 2020), and natural (data manipulation; Hullemann, Schupfer and Mauch 2017) sciences. However, few studies have used the NBL in exposure sciences. Furthermore, to the best of the author's knowledge, there have been very few applications of the NBL to determine the potential presence of underlying driving forces, such as homeostasis or hormesis, instead of its use in identifying data manipulation.

As described by Benford and pointed out by Brown (2005), some data will inherently violate the distribution, namely, (1) data that are confined by upper or lower limits, (2) experimental designs that only allow for only a few discrete values, (3) in systems where processes are present that influence the equilibrium or regulation to a value. Most fields have used the law to detect manipulated data and therefore this practice could result in falsely labeling data sets as manipulated when in actuality they are being driven by underlying mechanisms such as toxicokinetic and toxicodynamic (TK-TD) factors, seasonality influences, etc. as suggested by point (3).

We applied NBL to five different exposure data sets (Table 1) with the aim of understanding how TK-TD of essential, beneficial, and nonessential elements may influence conformity to the Newcomb-Benford distribution (NBD). We hypothesize that essential and beneficial elements will violate the NBD due to underlying physiological mechanisms that favor uptake and retention under normal conditions (observational studies), while non-essential elements will follow the NBD. We further hypothesize that, in experimental data sets (MICE) involving administration of elements, the data will violate the NBD due to the unnatural nature of dosing studies that manipulate the normal balance.

Data sets used in the present investigation were generated from various projects (Pino et al 2012, 2018; Alimotiet al n.d.; National Statistics Office of Georgia 2019; Ruggieri, Alimonti and Bocca 2016). The methods for measurement were based on Sector Field Inductively Coupled Plasma Mass Spectrometry (SF-ICP-MS) and followed the methods of Ruggieri,Alimonti, and Bocca (2016).

Data were processed using the 'BenfordTests ' package in R (v1.2.0; Joenssen 2015). Preliminary analyses assessed the applicability of various methods within the 'BenfordTests' package to the adult PROBE (PROBE_1) data set to determine the most suitable test for comparisons. Tests assessed included: Pearson's Chi-squared Goodness-of-Fit Test, Euclidean Distance Test, A Hotelling T-square Type Test, Joenssen's JP-square Test, Kolmogorov-Smirnov Test, Chebyshev Distance Test (maximum norm), Judge-Schechter Mean Deviation Test, and Freedman-Watson U-square Test.

Findings on the test comparisons suggest that most (7 of 8) tests are highly sensitive to small deviations from the BD; all but one test resulted in violations of the NBD ($p < 0.00001$). The least sensitive test was identified (Judge-Schechter Mean Deviation Test; Judge and Schecter 2009) and was selected to allow for quantifiable comparisons between violations of data sets.

All data were analyzed using the Judge-Schechter Mean Deviation Test for Benford's Law. This method reduces data points to a specified number of significant digits and then performs a goodness-of-fit test between the measured deviation means of first digit distributions and the theoretical ideal distribution of the NBD. Assumptions of this equation are that values should be continuous and that digits should be chosen so that the number of significant digits is not influenced by previous rounding. Further information can be found in the 'BenfordTests' package.

The J-S Mean Deviation Test requires that data are not confined by previous rounding; therefore, measurements found to be at or below the limit of detection (LoD) for each element and analytical method were removed from the data sets before analysis. Mean digit tests were not confined by digit number restraints and were performed using asymptotic analysis, as defined by the package. Data visualizations are provided as relative frequencies with 95% confidence intervals. Datasets for PROBE_1, PROBE_2, BURLO, and UNICEF were analyzed in three groups split by elemental classifications defined in Table 1 and also all together (All). According to the hypothesis, when all data points for all of the measured elements are combined, they should follow the NBD. Once divided into their respective groupings (essential, beneficial, and non-essential), we hypothesize that the data should violate, potentially violate, and follow the NBD, respectively.

Across the datasets, blood appeared to be the most reliable matrix to assess the current hypotheses. Urine and hair samples did not appear to display any accordance with the hypothesis; however more datasets are required to confirm this hypothesis (data not shown). This outcome could be due to the non-dynamic nature of these sample types compared to blood. After consideration of the biology of the sample types, more thorough analyses were conducted for the blood (PROBE_1, PROBE_2, UNICEF). After analysis of the PROBE_1 data (adult), it was observed that in All and Non-essential groupings, the data closely reflected a NBD ($p = 0.202$ and $p = 0.605$, respectively). Alternately, the essential and beneficial elements violated the NBD, as expected ($p < 0.00001$ for both). However, there appeared to be a non-random frequency distribution in the Essential and Beneficial results; the resulting trend of the data for these two sets appeared to follow a rotated sigmoidal curve, where the first digit of 1 resulted in a relatively high frequency, followed by a drop for digits 2 to 5, and ending with a slight increase and levelling off of frequencies from 6 to 9.

The PROBE_2 (adolescent) Essential data demonstrated a similar rotated sigmoidal curve and violation of the NBD ($p = 0.000987$). Conversely to PROBE_1 data, the remaining groupings of All, Non-essential, and Beneficial did not appear to demonstrate either random or fully Benford distributions ($p < 0.00001$ for each). For UNICEF data, All data followed the NBD ($p = 0.945$) closely, while Beneficial data depicted the previously described rotated sigmoidal curve ($p = 0.00366$), and the Non-essential data appeared not to follow either the NBD or another predictable trend ($p < 0.00001$). It could be noted that the differences in metabolic activity and energy demands between growing children/adolescents and adults could be the reason for the differences in these three data sets, particularly when discussing the unclear conformance of the non-essential elements in the children/adolescents. Other differences could be due to the small sample sizes and the uneven amount of elements between the groups.

When adherence to the NBD was compared within the blood biomonitoring data and between data sets (PROBE_1, PROBE_2, and UNICEF), two of three data sets followed the NBD in cases of nonessential elements, as was expected in the previous hypotheses. In the case of beneficial elements, there was inconsistent adherence with the three characterized curve shapes ie each depicted a different shape. However, both essential elemental groups analyzed (only for PROBE_1 and PROBE_2 data) demonstrated what appeared to be a characteristic rotated sigmoidal curve, which could be indicative of an underlying mechanism and could indicate active uptake of an element. This type of visual inspection could be indicative of a difference between falsely manipulated data and data with an underlying influencing mechanism. Previous research has identified any violation of the NBD as indicative of data manipulation. The use of violation of the NBL to determine manipulated data may be misunderstood and requires the consideration of whether the data in question is subject to an underlying driving force in the study system/subject.

173an Deepcdp: Deep Learning Charge Density Prediction

Siddarth Achar, Leonardo Bernasconi, Karl Johnson

Many applications in materials chemistry require information about charges and electron densities in materials and how they respond dynamically to chemical reactions, ion transport, etc. First-principles techniques such as density functional theory (DFT) provides both information about the atom positions and electron densities. This has allowed extensive research at the electron level and has enabled major developments in the understanding of materials chemistry. However, DFT can only be used for relatively small system sizes and short dynamic time scales because of the computational complexity of the calculations. To overcome this limitation, we have developed a *local* deep learning formalism for predicting charge densities only using atomic positions for molecules and condensed phase (periodic) systems. Our method, called DeepCDP (Deep-learning Charge Density Prediction), can predict electron densities for arbitrarily large systems by training it on DFT electron densities for small systems. We used *weighted*-Smooth Overlap of Atomic Orbitals (*w*-SOAP) to fingerprint atomic environments to their corresponding electron densities on a grid-point basis. We discuss techniques to determine optimized weighting and SOAP parameters that can effectively generate unique fingerprints depending on the material in consideration. Our DeepCDP formalism was tested for both molecular systems, like water, and periodic systems, like functionalized graphane. The models that were trained on small system sizes (<10 atoms) achieved electron density prediction accuracies of >99% on larger system sizes (>900 atoms). We demonstrate that DeepCDP achieves similar prediction accuracies for both charged and uncharged systems. We show how one can use the predicted densities for charged systems to perform interesting analysis such as computing the diffusion of charge centers over a large sheet of functionalized graphane. We make use of the ability of DeepCDP to predict densities locally to tackle intrinsic memory inefficiencies that occur during the evaluation phase, thus allowing us to perform rapid predictions for arbitrarily large system sizes.

173ao The Saturation Dependence of $BaSO_4$ Surface Precipitation Kinetics

Isaac Appelquist, Benaiah Anabaraonye, Philip Loldrup Fosbøl

Crystal growth on surfaces (crystallisation fouling) decreases the functionality of components in multiple industries. Fouling mitigation protocols include chemical dissolution or mechanical scrubbing, however, $BaSO_4$ remains a challenge, due to its low solubility in most common dissolution agents and its mechanical resilience. Therefore, to optimize $BaSO_4$ mitigation, it is important to understand its formation mechanism over a wide range of supersaturation conditions. A systematic study of $BaSO_4$ surface precipitation kinetics is missing in the literature.

In this work, we present a comprehensive investigation of $BaSO_4$ surface precipitation kinetics at saturation indices from 1.39 to 3.08. The saturation indices were calculated as $SI=\log10(a_{Ba^{2+}} * a_{SO4^{2+}}/K_{BaSO4})$

Deposition studies were performed in a custom-built once-through flow setup. Deposition rates were obtained through gravimetric analysis. High resolution X-ray microtomography was used to visualize the crystal formation. The morphology was quantified from texture parameters, including the spikiness (Pku), and the height of the tallest clusters. Furthermore, the distribution of crystal clusters was classified on a μm scale.

We observe a shift in growth modes, as the saturation indices are increased (see Figure 1). At low saturation indices (SI between 1.39 and 2.61), the growth occurs at a constant rate, while at higher saturation indices, the growth rate increases. These results provide new insights on crystallization fouling phenomena that are applicable in multiple industries.

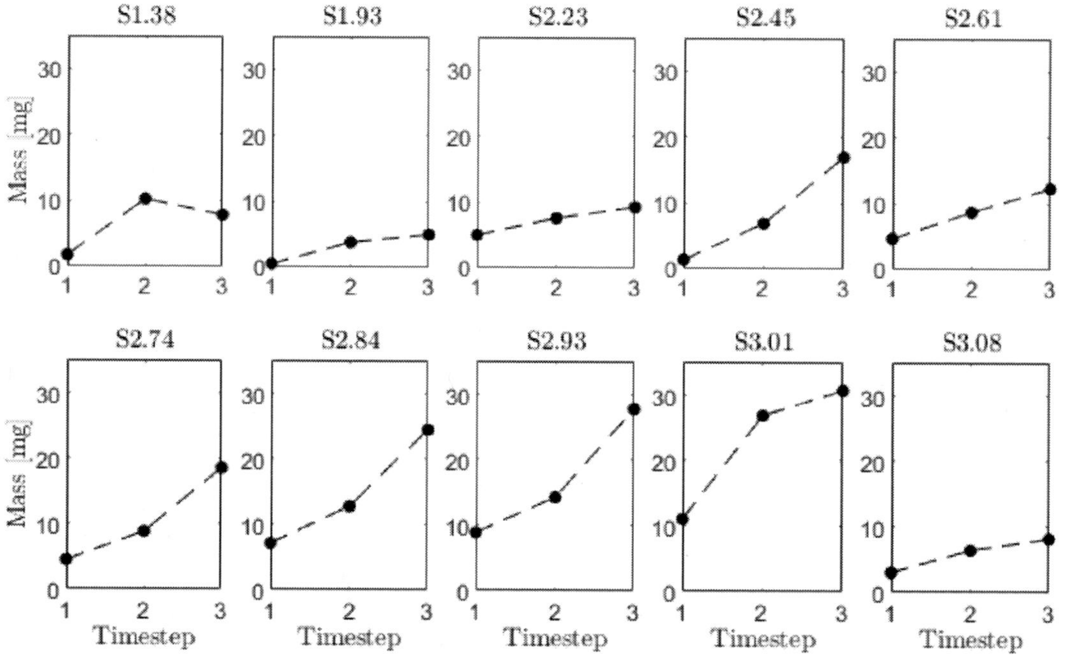

173ap High-Temperature Corrosion in Waste-to-Energy (WtE) Boilers and Cement Kiln

Janhvi Trivedi, Marco J. Castaldi

The combustion of the municipal solid waste in waste to energy plants leads to the formation of different types of chlorine and sulfur compounds which leads to corrosion on the surface of the boilers. Boiler corrosion is a crucial problem that leads to occasional shutdown of the plant which in turn increases the operating cost of the plant. Thus, to minimize boiler corrosion, we focus on studying the role of such compounds by carrying out laboratory tests using different boiler materials used in the industry. The laboratory tests are done to study the mass loss and corrosion rates of the materials when exposed to different environments and to select a material that can minimize corrosion in WtE plants.

For the WtE corrosion tests, Ni-based alloys with different compositions are used, and in cement kiln tests Fe-Cr-Ni-based alloys with various compositions are used. The results expected from these tests are that the mass loss in the Ni-based alloys and Fe-Cr-Ni-based alloys is decreased due to the formation of a more protective and passive oxide layer. The formation of such an oxide layer can minimize corrosion of boilers.

Three different boiler materials are used to study the high-temperature corrosion in WtE plant boilers and cement kilns. The materials used for the WtE experiments are Colmonoy 494 (0.42%C, 29.5%Cr, 3.48%Si, 1.2%Fe, 2.35%B, 58.2%Ni, 0.02%W, 0.01%Co, and 4.79%Mo), SP99 (0.76%C, 14.5%Cr, 3.53%Si, 13.8%W, 3.48%B, 1%Mo, 0.016%Co, 4.8%Fe and 54.2%Ni), SP686 (0.24%C, 8.9%Si, 22.1%Cr, 59.2%Ni, 7.26%W, 0.912%B, 0.014%Co, 2.8%Fe, and 5.57%Mo). The experimental materials used for the cement kiln tests are labeled as J (58%Fe, 25%Cr, 13%Ni, 1%Al, 1.2%Mn, and 1.1%Si), P (60.9%Fe, 22.6%Cr, 4.3%Ni, 0.5%Al, 10.4%Mn, and 1.3%Si) and F (63.9%Fe, 26.1%Cr, 5.5%Ni, 0.9%Al, 1.9%Mn, and 1.7%Si). The temperature range for these tests is between 400°C to 1000°C. The gas mixture used for the test is the WtE flue gas mixture and the cement kiln test is carried out in the oxidizing environment which includes CO_2 (15%), O_2 (5%), H_2O (10%), and N_2 (70%). These tests are conducted for 100-200 hours.

This presentation will show comparative results for several conditions and discuss possible reaction sequences leading to the observed corrosion rates.

173aq A KMC Based Tool to Understand the Chemical Recycling of Polyurethanes

SriBala Goruganttu, Matthew W. Coile, Obonetse Kebadireng, Linda Broadbelt

Polyurethanes (PUs) are the sixth most highly produced polymers globally and find a variety of applications like mattresses, insulation foams, automobile parts, etc. Traditionally, polyurethanes are formed by the addition reaction between diisocyanate and diol monomers. On the one hand, the raw materials used to produce linear and cross-linked PUs pose serious health hazards and environmental risks. On the other hand, conventional PUs offer limited recyclability and are incinerated, releasing toxic chemicals into the atmosphere. Solvolysis techniques like glycolysis and methanolysis have the potential to recover the monomers efficiently. Therefore, it is imperative to understand the chemistry of the PU solvolysis process at various reaction conditions to achieve higher monomer recovery rates.

The existing kinetic studies provide little to no information on the influence of reaction conditions, polymer structure, or molecular sequences on monomer recovery. These models generally track lumped polymeric species as a function of temperature or time. Tracking specific chain sequences and changes in the morphological aspects is essential as PUs consist of two monomers arranged randomly. Kinetic Monte Carlo (KMC) based models can help address this issue by explicitly tracking the sequences of all the polymeric chain species.

The current study aims at developing a KMC framework to unravel the depolymerization pathways of linear PUs. An in-house KMC model developed to generate linear PU chain sequences is used to initialize the polymer chains (hard and soft segments). Further, a set of reaction families are defined to track the depolymerization as single events. The rate parameters for solvolysis were optimized to fit the molecular weight distributions. The recovered monomer yields were validated against the experimental data available in the literature. Finally, a reaction map was plotted to understand the change in reaction dynamics – random vs. chain-end scissions – with solvent concentration, time, and temperature.

173ar Characterisation of the Crystal Polymorphs of Polyvinylidene Fluoride Using FTIR, Raman and DSC Measurements

Shubham Mireja, Devang Khakhar

Polyvinylidene fluoride (PVDF) is a polymer of considerable practical interest because of its amazing piezo and pyroelectric properties, in addition to it being chemically inert and mechanically strong. It is a semicrystalline polymer existing in five known crystalline polymorphs namely α, β, δ, γ and η, of which α, β, and γ are the more commonly encountered polymorphs. α form is non-polar whereas β and γ are the polar forms. β form having the highest polarisation density is the most desired form for ferroelectric applications. It is important to calculate crystallinity (X_C) and individual polymorph fraction with high accuracy. Some IR absorption peak positions are still ambiguous for the polymorphs and there have been multiple instances of confusion, especially between β and γ. Dynamic scanning calorimetry (DSC) can be used for finding crystallinity, but only for the PVDF sample containing α polymorph alone since for the other phases, the value of the specific enthalpy of melting, H_o has not been found yet. Peak positions can be more reliably assigned if spectra for pure crystalline polymorphs and the amorphous phase can be obtained. In this work, PVDF was processed into samples having predominantly α, β, and γ crystalline phases. These samples were analysed with FTIR, Raman spectroscopy, and DSC. For better qualitative and quantitative analysis of the phases, the spectra were fitted with the Gaussian peaks. Peaks exclusive to the amorphous phase were identified and used for the amorphous phase quantification. For the quantification of individual crystalline polymorphs in the sample, IR absorption coefficients for their respective peaks were calculated. Thickness of the samples was also calculated using FTIR spectra. DSC was used together with FTIR data to obtain the values of H_o for the α, β, and γ polymorphs and the values for H_o of β and γ were found not to be equal to that of α were found not to be equal as opposed to the assumption in the literature.

Construction of Magnetic Reactor with Ultrasonic Treatment for Rapid Clarification of Turbid Metamorphic Diesel Oil

Mubarak Azeez

Based on the previous study of this experimental work, it was found that 50 mL of modified turbid diesel oil was pretreated with 20 kHz and 80 W power ultrasonic at room temperature for 2 min, and then clarified by 60 mT magnetic field under the temperature of 50 °C, and 800 r/min stirring speed for 10 min. The magnetic field distribution was simulated with ANSYS software. Methylene blue dye channel in ultrasonic sound field test were utilized to obtain design parameters of the combined magnetic/ultrasonic treatment of turbid diesel oil. A set of reactor with 0-100mL/min controlled flow with adjustable temperature from room temperature ~ 70 °C was set up. Turbidity diesel coil pipe was constructed as well for diesel oil flow in ultrasonic device and a total number of 34 Z-shaped pipelines with the total length of 48 m. Two NdFeB permanent magnets are brought together in planes of the largest surface area and placed in groups of 5 mm apart. Using the above-mentioned ultrasonic/magnetic combination processing device, the modified turbid diesel oil was first subjected to ultrasonic wave of 20 kHz, pretreated at 80 W and 50°C for 2 min, and then magnetically treated at 80 mL/min and 50° C, which then reduced the turbidity of the diesel oil to less then 2.5% after experimental analysis of the results.

INTRODUCTION

Application of magnetic processing oil technology and ultrasonic technology was used on turbid metamorphic diesel oil to explore the change in viscosity after ultrasonic solidification of turbidity and metamorphic diesel oil. In this study, it takes long time to prepare for turbid degraded diesel according to the traditional method (like exposure to sun). A fresh 0# diesel fuel was modified by introducing oxygen to accelerate its deterioration in other to speed up the progress of the experiment shown in **Fig. 1.1** and **Fig.1.2**. Study on the effect of magnetic field combined method on the apparent viscosity of metamorphic diesel oil to provide reference for the industrial application of ultrasonic magnetic field treatment to reduce the apparent viscosity of turbid metamorphic diesel oil. The following chapters in this paper explain and explore the process of the clarification of the turbid metamorphic diesel oil by applying a constructed magnetic reactor with support of ultrasonic technology

Hazard of using turbid metamorphic diesel

Due to the small clearance between the injection pump's plunger, outlet valve, fuel injector's coupling, and diesel lubrication during operation, the use of diesel with more dust and impurities will accelerate the three precision parts. It affect the mechanical wear, plunger parts that greatly reduce the sealing performance and the effective oil supply stroke also reduced than normal so that the oil supply from the oil pump rapidly decrease. The lower the rotational speed, the lower the throttling effect, the greater the leakage the amount of oil supply. The more pronounced is the reduction which is detrimental to the start-up and idle operation of diesel engines. In addition the smoke coming out of such diesel also increases the CO2 that affect the climate change and environment.

By apply the designed reactor this drastically reduce the breaking of engines and it lower the carbon released to the environment. Research and laboratory analysis also shows that diesel fuel renewed using magnetic reactor increases the power of the engine to perform far better than fresh diesel.

173as Hollow Fiber Membranes for Evaporative Cooling

Andrew Lin, Dibakar Bhattacharyya

Hollow fiber membranes are widely used as membrane contactors, for gas separations, and for filtration applications. Insight into the relationship between membrane parameters such as porosity, tortuosity, pore diameter, and membrane thickness can be used for the development of mathematical models for performance characteristics. Information is presented on a series of vacuum membrane distillation experiments conducted on a highly hydrophobic polypropylene hollow fiber membrane module. This data investigates the relationship between membrane parameters and the heat transfer that occurs throughout a module as driven by evaporation. Water fed through the lumen side of the polypropylene hollow fiber membrane module will evaporate through the pores of the membrane. Because of the nanometer pore size and hydrophobicity of the membrane, a liquid-vapor interface is suspended across each of the pores by the surface tension of the water. During experimental trials, heat transfer from the lumen water measured through the variation of flow rate and shell side pressure. Application of this data to a series of fluid and mass transfer models was used to evaluate heat rejection values. Future applications of this model will provide valuable information on heat transfer phenomena in hollow fiber membrane-based separations.

173at Integrated production of chemicals and fuels in the pulpindustry: techno-economic and environmental analysis ofblack liquor gasification-based processes

Meire Ellen Gorete Ribeiro Domingos, Daniel Florez-Orrego, Silvio de Oliveira Junior, François Maréchal, Moises Santos

The pulp and paper sector is classified as an energy intensive activity and several efforts are being made to mitigate its atmospheric emissions, improve the recovery of residual heat and capitalize on its byproducts. The black liquor (BL) is a byproduct of the kraft pulping process, which contains more than half of the exergy content in the total woody biomass fed to the digester, thus, representing a key supply of renewable energy to the pulping process. The BL can be gasified and the syngas can be used to generate electricity or produce chemicals and biofuels, through the broader biorefinery concept.

In this work, the conventional scenario of the BL use (i.e. concentration and combustion) is compared with the BL upgrading gasification process for different chemicals and fuels (namely, hydrogen, synthetic natural gas, methanol, dimethyl ether, ammonia, urea and nitric acid) production, in terms of economics, exergy efficiency and environmental impact. A combined energy integration and exergy analysis is used to identify the potential improvements that may remain hidden to the energy analysis alone, namely, the determination and mitigation of the process irreversibility. The chemical processes synthesis, modelling and simulation are performed by using Aspen Plus® software. Meanwhile, the determination of the minimum energy requirements and the solution of the energy integration problem is handled by the OSMOSE Lua platform. A sensitivity analysis of the variation of the INPV as a function of the carbon taxation (0-100 EUR/tCO2) and the interest rate (0-21%) is performed. Firstly, it is assumed that the present values of the feedstock and the products are set as constant. Next, the incremental financial analysis incorporates the uncertainty related to the acquisition and selling costs of the feedstock and fuels produced, when embedded in a volatile market, by using the Monte Carlo method.

As a result, the exergy efficiencies of the conventional setup and the kraft pulp and the integrated plants average 40% and 43%, respectively; whereas the overall emission balance varies from 1.97 to -0.79 tCO2/tPulp, respectively. In anticipation of future carbon taxation scenarios, the incremental economic analysis found that the fuels and chemicals production route with partial electricity import may economically outperform the conventional kraft pulp mill for moderate carbon taxations (30-100 EUR/tCO2), depending on the interest rate adopted. These results reinforce the relevance of the electricity import from the Brazilian mix for pushing upwards the share of renewable energy resources in the production of traditionally fossil-based chemicals and fuels, which could be an important decision variable in favor to the exploration of integrated biorefineries.

Keywords: Biorefineries, Process Design andamp; Development, Forest andamp; Plant Bioproducts, Biofuels (Energy), Efficiency (Energy)

Way to Reach Energy Independency

David Judbarovski

Being very productive principle inventor and investigator on a firm basis of my work in wide spectra engineering and economics, I am sure that energy and production have to be and are ready to be green andamp; cheap andamp; abundant andamp; inexhaustible, while all materials and parts and tools for that purpose can be recyclable.

Regarding to fossil fuels, they can and must be substituted by artificial oil and gas, can be competitive if primary energy to produce them being cheaper 1.0 cent/kWh.

Below I show some ways to reach such cost target.

To produce a barrel of artificial oil of refined quality we need

(44 * 1400 + 6 * 50,000) * 0.159 * 0.85 /14 = 3,500 kWh /boe (barrel of oil equivalent) as OPEX

Here:

44 – CO_2 molecular mass;

1400 (kWh) - electric energy for 1.0 ton CO_2 production (see Note 1)

6 – molecular mass of 3 H_2;

50,000 (kWh) –electric energy for 1.0 H_2 ton production by the water splitting by concentrated rays of standard magnetrons;

0.159 (m^3) – barrel;

0.85 (ton/m^3) – density of oil of refined quality

14 – (CH_2) molecular mass

Moreover, existing and well known technologies allow halving demand in oil and gas by water vapor reforming them in hydrogen in very small flow rate, so such hydrogen consume is hazard free and allows twice more energy effective electric generator being hydrogen-air fuel cells, being compact and cheap device if mass produced vs. conventional direct oil and gas usage for energy needs.

Different inventions promise primary renewable electricity cheaper USD 0.01/kWh each at 5 years payback and then practically free.
They are.
1. Solar thermal heterodyne. 2020. It allows a rectenna to be feasible at present level of rectifiers. About 0.6 cent/kWh electricity (see Note 2).

2. Sailboats with hydro-generator. 2017. Its electricity is transformed on-board in hydrogen, and then in more transportable fuels at floating nearby terminals. About 0.35 cent/kWh electricity.

(Fresh look on Wind Power Generation - Upgraded, 25/02/2019, David Judbarovski,

2019 AIChE Annual Meeting (ISBN: 978-0-8169-1112-7), https://www.aiche.org/proceedings/people/david-judbarovski }

Breeding investment allows by credit for a first power plant in five steps to invest and build some tens analogous power plants. .

So OPEX is negligible vs. CAPEX, can be full cost of artificial intermediary fuel for 100% off-grid electricity by 2.5 cent/kWh. For housing such autonomy can be provided by local fuel-in-electricity substations. Autonomous supply can be paralleled by much cheaper direct grid supply.

Note 1.

CO_2 can be USD 15.0/ton being my estimation with a good reserve, and being produced in quantity of 320 ton/m2/year in production ponds and by electricity cheaper than USD 0.01/kWh

(579f) CO_2 Mass Production from the Air Very Cheaply, 2020 Virtual AIChE Annual Meeting (ISBN: 978-0-8169-1114-1), https://www.aiche.org/proceedings/people/david-judbarovski

Note 2.

The said goal is achievable by solar power plants operated in snow-free seasons only and reserved in a form of artificial fuel, mainly for snow seasons.

Solar electro-thermal power plant of my design, it is original solar concentrating system with thermal heterodyne in focal spot being motionless and serving to convert high temperature heat in electricity with energy efficiency about 67%, while mechanical part of the concentrator is about USD 15.0/kWp. It is Antenna source of electricity being rectenna with rectifiers about 2.5 THz or so.

I suppose 7 months (since April to October) in a years as operational time, with 12 hours daylight with cloudiness 50%, 12 h * 200 d * 0.5 = 1200 kWh/m2. Supposing Rectenna to be some times more expensive than PV cell and to be USD 1.0/W and concentration ratio 100 : 1, it is USD 10/ kW. Being summarized, it is (15 $ + 10$) / (1200 kWh * 5 years payback) / 0.67 efficiency = 0.6 cents/ kWh electricity.

An MILP Based Soft Sensor Design Algorithm

Hasan Sildir, Sahin Sarrafi, Onur Can Boy

Significant amount of online and offline data is collected from industrial and complex plants. The decision making based on such data requires selection and management of statistically contributing and easily measurable variables for the prediction of quality related variables which are obtained in relatively lower frequency and with significant delays due to laboratory issues. There are many methods to develop mathematical models, named as soft sensors, which deliver online predictions for such laboratory measurements. The accuracy and practical aspects of these soft sensors are important concerns due to product quality issues and plant maintenance.

This study focuses on the development of a mixed integer nonlinear programming (MINLP) problem to address the selection of a representative set of online measurements in addition to their lags. Those lags are calculated for each online measurement and delivers a better soft sensor performance since it explicitly introduces the time required for the mass transport from each sensor to the sample collection point. Moreover, the approach introduces binary variables into the formulation, in addition to traditional continuous parameters, to account for the existence of online measurements. The formulation is further tightened with linking constraints to ensure the zero-coefficients for the non-existent variables and narrow the search space for the optimization algorithm. The problem formulation is further modified to result in a mixed integer linear problem (MILP) by piece-wise representations of nonlinear expressions. Such an approach enables the selection of best theoretical subset of online measurements to deliver minimum prediction error. The overall method is flexible to different processes and problem related specific needs as more constraints are introduced to the formulation.

The methodology is implemented on an actual industrial distillation plant for the prediction of kerosene boiling point related quality variable and delivered approximately 25% better online performance compared to existing soft sensor. In addition, several optimization algorithms are compared in terms of CPU and accuracy for the solution of corresponding optimization problems.

Upcycling Waste Plastic to Graphitic Carbons By Graphene Oxide Addition

Akshay Gharpure, Randall Vander Wal

In the United States, approximately 37 million tons of plastic are used every year. More than 70% of the single use plastic waste is sent to landfills. This presents a huge opportunity to upcycle the plastic waste into high value carbon materials. Concurrently a big surge in demand for carbon materials is expected from energy storage and transportation applications. Flake graphite production would need to double by 2025 just to meet Lithium-ion battery (LiB) manufacturing capacity currently under construction.

When heated, plastics typically crack into light gases through chain unzipping and β-bond scission. This translates into low carbon yield and non-graphitizable residual carbon. We demonstrate an innovative approach of employing GO as templating agent and closed reactor carbonization under autogenic pressure to increase yield and graphitic quality of carbons. Oxygen functional groups on GO provide the necessary stabilization while the sp^2 framework would serve as a template guiding reconstruction of polymer chains into graphitic material. The latter postulated templating effect could realize significant cost and energy savings by enabling graphitic structure at lower temperatures. Upcycling overabundant plastic waste into graphitic carbons supports the renewable energy transition while reducing pollution, cost and CO_2 emissions.

Four commercially recycled plastics and their composites with GO have been used to evaluate comparative yield and graphitic quality of the obtained carbon materials. GO provides excellent stabilization through the radical sites formed by the leaving groups. The key advantage of GO as additive is that it nets a substantial increase in carbon yields, nearly 300% in some plastics. There is a trade-off between carbon yield and graphitic quality in GO/plastic composites. After graphitization, graphitic quality of GO composites is slightly lower than all pure plastics (except HDPE) potentially because of cross-linking by reactive radical sites on GO which is responsible for the yield increase.

A Fault Detection Framework for a Class of Unknown Nonlinear Spatio-Temporal Systems from a Systems Modeling Perspective

Yun Feng, Yaonan Wang, Yazhi Zhang

Spatio-temporal systems contain but are not limited to flexible mechanical systems, battery thermal processes, and chemical processes. They can be concluded into distributed parameter systems (DPSs) whose dynamical behaviors are modeled by partial differential equations (PDEs) in general. Fault diagnosis of DPSs is vital to the safe and efficient operation of these systems while it has not been well investigated in the literature. For example, the battery thermal faults which lead to thermal runaway and even fire are the major causes that restrict the safety operation of electric vehicles and hybrid electric vehicles. Fault detection for distributed parameter systems reported so far are model-based in general and the performance heavily relies on the prior known model information. This restricts the usability of these methods in industrial applications. In this paper, we make the first attempt to establish a unified framework that contains both on-line systems modeling and fault detection of unknown high-dimensional DPSs. These two parts interact with each other in the sense that the systems modeling error is transformed into the residual signal for fault detection while the on-line modeling switches to off-line mode depending on the fault detection results. The high-dimensional DPSs are first decomposed into spatial features and temporal sequences. Then a receding-horizon scheme is applied for the temporal dynamics learning and the residual signal is converted by the temporal validation error. Experiments on sensor faults diagnosis for the thermal process of a 2-D battery cell are provided for method validation.

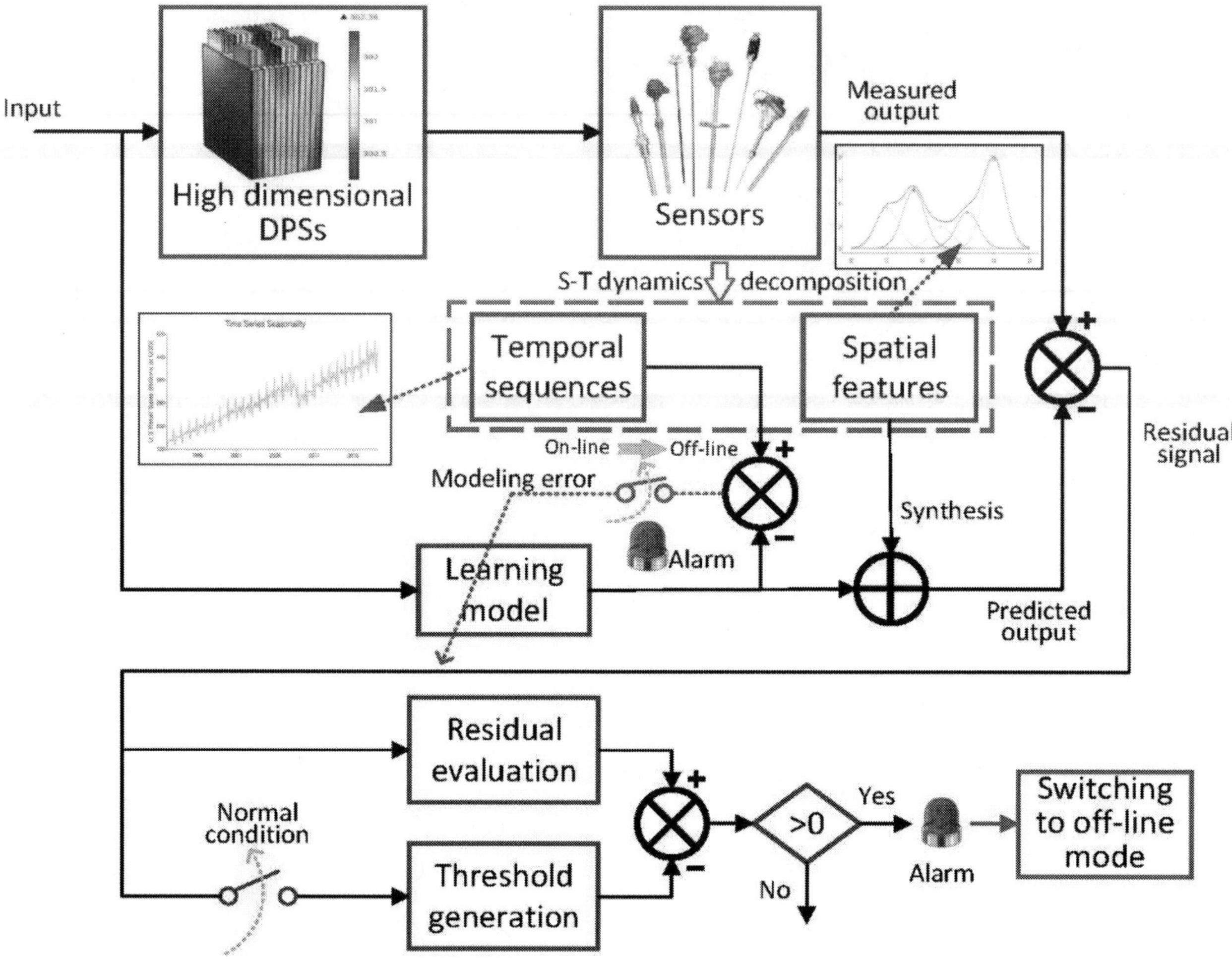

Input
High dimensional DPSs
Sensors
Measured output
S-T dynamics
decomposition
Temporal sequences
Spatial features
Time Series Seasonality
On-line
Off-line
Modeling error
Alarm
Learning model
Synthesis
Residual signal
Predicted output
Residual evaluation
Normal condition
Threshold generation
>0
Yes
No
Alarm
Switching to off-line mode

Creating the First Combined, on-Site and Remote, Cheme-Sports^{TM} Competition

Matthew B. Garvey, Daniel C. Vanderklay, Michael A. Lampasona, Donald C. Glaser, Robert G. Bozic

ChemE-Sports^{TM}, a worldwide chemical engineering optimization and safety competition, is now capable of combining on-site and remote competitors. Inspired by the success of ChemE-Sports^{TM} from 2019 to 2020, an academic and industry team of a faculty member at Columbia University Chemical Engineering and PetroSkills - Simulation Solutions developed a new method of using simulators for a simultaneous on-site and remote competition at the 2021 AIChE Annual Meeting in Boston, MA. This newly introduced capability can expand chemical engineering competitions for the future. The competition was created around the safety (i.e. management of change, troubleshooting, etc.) and optimization of an Amine Gas Treating System simulation owned by PetroSkills- Simulation Solutions. A competitive team bracket featuring 25 teams from 8 countries was developed for the event during the AIChE Student Conference on Sunday 7 November 2021. The results and lessons learned of the 2021 competition are presented here. This presentation will focus on the competition and modifications for combined, on-site and remote participation from student groups around the world.

Creating an Auto-Encoder Based Predictive Maintenance Tool for Offshore Annulus Wells

Akhilesh Jain

A key component of well integrity is annular integrity. Much of the focus on this has been on establishing maximum and minimum pressure limits and designing envelopes under various well conditions encountered during well construction and subsequent production and injection operations. Many operators have established systems for operating wells within this design envelope to monitor for pressure excursions. However, abnormal annulus pressure behavior within the design envelope could be overlooked using a system that relies on limit monitoring and excursions.

We propose a modeling workflow that combines novel deep learning techniques with statistical analysis to create online models which predict potential asset failures and alert on abnormal behavior such as abrupt pressure build up in producer and water injection annulus wells. The model uses autoencoder deep neural network architecture to learn the behavior of the wells during normal operating periods and generates alerts when it encounters new or abnormal behavior.

The autoencoder architecture outputs a risk score aggregated over the residuals from all input features. Sequential Probability Ratio Test (SPRT) is performed on the risk score to determine abnormal regime during operation to raise alerts. These alerts can be used for root cause analysis based on the top contributors to the risk score. In our approach, we use feature thresholds as filters to determine normal operating periods for training the model. To simulate live conditions during model training, the historical time series data is divided into training and prediction windows. The model is trained on each training window and risk scores created for the prediction window using a sliding window technique. To find the optimum model, a grid search is performed over a wide distribution of autoencoder and SPRT hyper-parameters. The models are scored based on recall, precision and lead time provided before a failure.

We demonstrate this workflow using annulus pressure, downhole pressure, upstream choke pressure and upstream choke temperature as input to the model. The model does not require the physical properties of a well but uses historic well data lending itself to be applicable to already existing well stock. Next, we demonstrate using engineered features and synthesized data to efficiently train and score the models. During our experiments, we have explored several engineered features across multiple platforms and found that the correct set of engineered features can deliver a model that accurately alerts on asset abnormalities and potential failures.

Our approach combines the strengths of deep learning techniques, statistical analytics and subject matter expertise to provide a framework that has demonstrated efficient scaling across multiple assets and sites and has potential application on a variety of oil and gas equipment.

Simulation of a Rotating Ring Disk Electrode System Incorporating Ionic Migration

Sirshendu Guha

Previous research works using rotating ring disk electrodes (RRDE) mostly were restricted to situations where ionic migration had been neglected on account of presence of excessive supporting electrolytes in the solutions. However, there are many situations where supporting electrolytes are not present or its concentrations are negligibly small. Examples of such situations are anodic dissolution processes, metal deposition processes etc. Migration of an ionic species plays very important role in these situations with regard to transport of the species to any electrode from the bulk solution and as a result currents at the electrodes change substantially in presence of migration.

In this work, effects of ionic migration have been investigated for a rotating ring disk electrode (RRDE) system in which the physical situation is described in the following paragraph [1].

The electrolytic system is a binary aqueous salt solution of copper sulfate ($CuSO_4$) in which sulfuric acid (H_2SO_4) is used as supporting electrolyte. The electrolytic system portraying presence of migration case is copper sulfate ($CuSO_4$) solution of 0.1(M) concentration (Cu^{+2} concentration in bulk = 0.1(M)) with very little concentration (0.01 (M)) of supporting electrolyte (H_2SO_4, H^+ concentration in bulk = 0.02 (M)). In contrast, the electrolytic system describing absence of migration case had been copper sulfate ($CuSO_4$) solution of 0.1(M) concentration (Cu^{+2} concentration in bulk = 0.1(M)) with very large concentration (10 (M)) of supporting electrolyte (H_2SO_4, H^+ concentration in bulk = 20 (M)). The electrode reactions at the disk and at the ring involve transfer of single electron at the electrodes which can be depicted as,

$Cu^{+2} + e = Cu^{+1}$ at the disk

$Cu^{+1} = Cu^{+2} + e$ at the ring

SO_4^{2-} is the conjugate ion whose bulk concentration is 10.1 (M) in case of absence of migration and 0.11(M) in presence of migration case [1].

These effects were determined (at limiting current condition) by numerically solving the dimensionless ordinary and partial differential equations (equations for disk, gap and ring regions of the RRDE in non-dimensional forms) which describe mass transfer in absence of ionic migration vis-à-vis in presence of ionic migration. These equations were solved using explicit finite difference numerical method on a high speed computer. In order to address stability constraints of explicit finite difference numerical technique, the step sizes in normal (x) and radial (r) directions had to be optimized as $\Delta x = 0.046$ and $\Delta r = 0.005$ respectively. Collection efficiency of a RRDE which is defined as ratio of quantity of material detected at the ring to the amount of material generated at the disk is also evaluated for various electrode geometries and compared for presence and absence of ionic migration cases [1].

The numerical experimentation results indicate that in case of presence of migration, the collection efficiency of the electrolytic system described above which involve rotating ring disk electrode increases for a given electrode geometry (disk radius, gap thickness and ring thickness) as compared to the collection efficiency in absence of migration case. Simulation results also indicated that collection efficiency of any RRDE system is dependent RRDE geometrical parameters (ratio of outer gap radius (r_2) to disk radius (r_1), ratio of outer ring radius (r_3) to outer gap radius (r_2)) [1]

Numerical results also show that in case of presence of migration, the concentration boundary layer thickness for the limiting reactant (Cu^{+2}) reduces by about 30% when compared with this thickness of the concentration boundary layer existing in absence of migration case [1].

It is also found that ionic migration remains critically relevant for supporting electrolyte (H_2SO_4) concentration as high as 0.1 (M) [1].

Literature Cited

- Guha, S. Simulation studies on a rotating ring disk electrode system: Role of supporting electrolyte in determination of relevance of ionic migration (2013). ***AIChE Journal.*** 59(4), 1390–1399. doi: 1002/aic.13903

Effect of a Slope Sign Loss Term on Prediction Accuracy and Consistency of a Machine-Learning Model

Jaewook Lee, Subeen Lee, Yoonae Ko, Jeongmin Kwak, Jay H. Lee

As more and more data are collected in the process industry, the importance of data-driven modeling is increasing. Since the detailed first principles relating the process variables are often unknown, either partially or entirely, use of the grey-box or black-box model is the industrial norm. Lately, machine learning-based models like the Recurrent Neural Network (RNN), Long-Short Term Memory (LSTM), and Gated Recurrent Unit (GRU) have received attention as an alternative to the traditional statistical models such as Auto-Regressive Integrated Moving Average (ARIMA) in capturing the patterns seen in time-series process data.

There are two main paths to increasing the prediction accuracy of a black-box model. The first path is to modify the structure of the baseline model, which is especially widely used in the computing field. An exemplary approach along this path is an attempt called attention1. This approach has shown strength in various fields, but in general, a modified model ends up being more complicated than the baseline model. Therefore, the main weakness lies in the time needed for finding a robust model structure and training it with new data.

The second path is to change the objective function to make the training fit better to the model's end purpose. This is simpler as the structure of the existing baseline model can be utilized and the previous training provides a good starting point for searching for the best hyperparameters. Representative methods include methods such as Ridge2 and Lasso3 using the l1-norm and l2-norm in addition to the Mean Squared Error Loss (MSELoss).

In the PSE field, use of a black-box model is prevalent for guiding how to change inputs to optimizing or controlling outputs. In this case, it becomes very important that the model is fitted to be consistent with the directions (e.g., sign of the slope) of the output responses. Slope sign, one of the main trend indicators for time-series data, can be an important aspect of the prediction accuracy. For example, a model that predicts the direction of output change in the wrong way would lead to control actions that are opposite to proper remedial changes, manifesting as closed-loop instability. The same is true in feedback-assisted optimization. For example, when one tries to optimize a given function using a gradient-based algorithm, the slope sign error would take the search further away from the local optimum leading to divergence. In such applications, it is of utmost importance that a data-driven model matches the sign of time trends of the output variables with high accuracy. On the other hand, since noise terms can easily reverse the slope sign of the output variables, it is important to consider their impact.

Motivated by this, we propose a new objective function called the *slope sign loss*. This function weighs the incorrect prediction of the slope sign through an additional penalty during training. The final objective function is a weighted sum of the MSELoss and the slope sign loss. The weights can be chosen to minimize Root Mean Squared Error (RMSE) and slope sign error by treating weight of the slope sign loss as a hyperparameter.

As a case study, 2 years of the transoceanic transport data of LNG shipment provided from Korea Shipbuilding andamp; Offshore Engineering (KSOE) Co. were used as a dataset. In addition to the data, a prediction model was constructed to predict the pressure of the tank and the simulated data from the model were added. Gated Recurrent Units (GRU) was trained as the prediction model. The raw data were originally sampled every minute. However, since internal pressure of the LNG tank does not change rapidly in response to a sudden change of input variables, voyage data was resampled to an hourly rate. Subsequently, 9 input variables were selected for their high correlations with the output variables based on Pearson's coefficient, Kendall's tau, Spearman's coefficient and our additional physical insights. Outliers were detected and removed with the IsolationForest algorithm. For filtering, the Savitzky-Golay filter was used since this filter showed an advantage over other filters such as ARIMA, EWMA in terms of filtering time and quality.

In general, the results of the prediction model do not vary significantly depending on the random seed that determines the initial state of the machine learning model. An exception was the case of process data for which the prediction results often changed depending on the initial state due to the insufficient quantity. A prediction model with consistent prediction results with different random seeds is said to possess random seed stability4. From this point of view, we have attempted to show that the prediction model fitted with the slope sign loss not only shows good accuracy in a particular random seed, but it also displays the desired trait of random seed stability.

To illustrate the effect of adding the slope sign loss to the training, three error metrics were examined including train loss, test loss, and slope sign errors. Those error metrics were extracted from a total of 1000 random seeds with 10000 epochs each. In order to examine the effect of adding the slope sign loss term on random seed stability, the standard deviations of the losses at 3000, 5000, 7000, and 10000 epochs were compared in addition to the mean of the error metrics at those epochs.

The results confirmed that the average train loss, test loss, and slope sign errors all decreased when the slope sign loss term was added to the objective function regardless of what weight was assigned. In particular, train loss and slope error tended to decrease gradually as the weight of the slope sign loss term increased. Standard deviations of the 3 types of losses also tended to decrease. Therefore, it may be said that random seed stability is improved through the use of the slope sign loss term. The case study results reveal that adding the slope sign loss term to the objective function gave a prediction model that not only followed the trend of raw data well, but also showed less variations in the prediction results with respect to different random seeds.

1. Qin, Yao, et al. 'A dual-stage attention-based recurrent neural network for time series prediction.' arXiv preprint arXiv:1704.02971 (2017).
2. Yu, Xinying, and Shie-Yui Liong. 'Forecasting of hydrologic time series with ridge regression in feature space.' Journal of Hydrology 332.3-4 (2007): 290-302.
3. Li, Jiahan, and Weiye Chen. 'Forecasting macroeconomic time series: LASSO-based approaches and their forecast combinations with dynamic factor models.' International Journal of Forecasting 30.4 (2014): 996-1015.
4. Madhyastha, Pranava, and Rishabh Jain. 'On model stability as a function of random seed.' arXiv preprint arXiv:1909.10447 (2019).

Development of Zero-Dimensional Model for Air Gap Membrane Distillation (AGMD) Desalination System

Hassan BaAqeel

Membrane distillation (MD) is a thermal separation process that employs chemical potential difference, essentially vapor pressure difference, between two fluids at different temperatures to drive the mass transfer separation across a hydrophobic membrane. The utilization of a hydrophobic micro-porous membrane reduces the phase separation space required, otherwise, in other thermal separation processes such as conventional evaporators. MD has many other reported advantages over other separation processes including lower operating temperature and higher tolerance to salt concentration and fouling . It has a subtle application in the field of desalination and wastewater treatment as well as the removal of heavy organics and metals.

There are many types of MD, each striving to strike a balance between heat efficiency, water flux, and capital expenditure. Direct contact membrane distillation (DCMD) is the simplest MD design with the least capital expenditure. However, it endures relatively higher heat loss, allegedly as high as 40%, which increase its operating cost. Air gap membrane distillation (AGMD) employs an air gap between the two fluids to minimize heat losses through conduction. Thus, the air gap adds a layer of mass transfer resistance reducing flux in the system. Vacuum membrane distillation (VMD) eliminates this resistance by introducing a vacuum gap instead of an air gap. This comes with the drawback of an increase capital and operating expenditure due to the added vacuum system. Other configurations are also proposed, such as filling the air gap with different materials or water, viz. water-gap MD (WGMD) or material-gap MD (MGMD).

MD has grasped immense research interest in recent years. The number of research activity on MD has almost doubled in the last decade between 2000 and 2010. The upsurge in MD research in early 1980s was due to breakthroughs in membrane manufacturing that improved the membrane characteristics and performance. It is worth noting that the highest researched configuration in MD is DCMD followed by AGMD and VMD. In the authors' opinion, the focus on DCMD is an attempt to develop a more fundamental understanding of the MD process and the available membrane performance. Hence, DCMD research has been very useful to establish the base for other configurations to evolve. Corroborating the authors prospect is the trend of commercializing MD configurations other than DCMD, such as AGMD in the case of MEMSTILL and vacuum enhanced AGMD in the case of MEMSYS.

Despite the heavy research on MD, the technology is, hitherto, economically challenged in the field of water desalination due to its low capital productivity and, costing uncertainty, operational immaturity. Operational immaturity refers to our understanding of the long-term operational performance and fouling of MD. The low capital productivity is due the relatively low flux which entails high CAPEX for any reasonable residential or industrial production level. One solution to overcome the economic challenge of the MD system is the integration with other desalination technologies, industrial processes, or renewable heat sources. For optimizing such integrated designs, performance and cost models for the relevant MD technology has to be established. Being the most researched configuration, DCMD attracted much of the work for heat and mass transfer modelling as well as cost estimation . Thus, the effort in other MD configurations such as AGMD is nowhere near exhaustive.

Models are generally categorized by the number of space dimensions they cover. Zero-dimensional (0-D) models evaluate heat and mass transfer based on the average conditions over the whole membrane module. On the other extreme, 2-D models (i.e. CFD) evaluates the heat and mass transfer across the 2-dimensional membrane surface and is computationally very expensive. Available AGMD models assume different level of dimensions. Amongst others, Chernyshov, Meindersma and de Haan has established a 2-D model for temperature and salt distribution in AGMD. Alklaibi and Lior focused their AGMD modeling work on a 2-D transport analysis of AGMD. For 1-D modeling of AGMD, two research work stands out by Guijt, Meindersma, Reith and De Haan and Alsaadi, Ghaffour, Li, Gray, Francis, Maab and Amy

In this work, a zero-dimensional (0-D) model has been developed for air gap membrane distillation (AGMD). This model is based on the decoupling of the heat and mass transfer mechanisms and empirical correlations to get around the iterative nature of one-dimensional (1-D) models. The model has been validated for AGMD design with a temperature difference (DT) of 5-30°C across the membrane, a range deemed the practical DT for any commercially viable AGMD designs. The validation is carried against both experimental data obtained from literature as well as rigorous 1-D model results for two commercially available membranes. The comparison of the predicated flux is carried out for ranges of air gap width and operating conditions. The model predictions are reasonably correlated with the experimental and rigorous results with an error within 5% at low DT that increases to 18% at large DT. The model is foreseen very useful in the optimization of large plant design systems integrating AGMD technology without making the model computationally very expensive. The applicability of the model is illustrated in the optimization of the MD technology selection for desalination using industrial waste heat. A case study has been presented in this paper to illustrate the applicability of the newly developed model in plant design optimization and cost.

Guiding Membrane Design and Development By Machine Learning

Sui Zhang

Machine learning has been a popular tool in the discovery and development of new materials. With proper algorithms, different models may be built to analyze the complex systems, identify the key features and guide material synthesis. In this presentation, we will introduce some of our preliminary efforts in applying machine learning for the fabrication of better membranes. First, Gradient Boosted Tree Models were trained to understand the material design and fabrication principles of thin film nanocomposite membranes for reverse osmosis. Second, literature data on metal organic framework based mixed matrix membranes for CO_2 separation were collected and trained by Random Forest Models. Based on model guidance, suitable MOFs were synthesized and membranes showed excellent performances which match well with model prediction. Transfer learning was also applied to save the resources. Third, artificial neural network was applied to reveal the formation mechanisms of nanofiltration membranes. Thereafter, membranes with high ion selectivity and permeance that surpass literature work were fabricated based on model guidance.

2D Membranes for Molecular Separations

Sui Zhang

2D materials offer the opportunity to overcome the permeability-selectivity tradeoff through in-plane nanopores and inter-plane slits. Single-layer graphene is the thinnest possible separation layer (atomic thickness) and the possibility of pore creation across graphene to allow size-based molecular sieving of different molecules. In particular, its excellent stability in organic solvents make it an ideal candidate for organic solvent nanofiltration. To enable effective molecular separation in organic solvents, two approaches have been demonstrated in our recent work, one is the direct transfer of graphene onto electrospun fibers, and the other is the integrated design of support to control the transport of molecules across the composite membrane. We have shown that the solvent permeance of graphene membrane can be two to three times higher than literature work. In addition, excellent stability in high-temperature, harsh organic solvent has also been observed. Then, laminar membranes based on 2D conjugated polymers and their application for gas separation will be discussed. With microporosity and interlayer spacing, a hydrogen permeability of $> 20{,}000$ barrer at a H_2/CH_4 selectivity of > 10 has been achieved, and excellent stability has been demonstrated over a 30-day test at 150 °C.

Bubble Column :Radial Distrubtion Study of Gas Hold-up Under Different Operating Conditions

Ghanim M. Alwan Sr.

Bubble columns BCs are multi-phase units where the gas phase is dispersed into a continuous liquid phase in two-phase columns . The simplest bubble column layout consists of a vertical cylinder without internals, in which the gas enters at the bottom— through a gas sparger (i.e., porous gas sparger, perforated plate, ring gas sparger, needle gas sparger or spider gas spargerand the liquid phase is supplied in batch mode or it may be led in either counter-currently or co-currently to the upward gas stream. The effects of gas velocity Ug(0.2-0.3 m/s) and system pressure P(0.1-1.0 Mpa) on gas hold up Hg and liquid hold up Hl were studied in a 0.162 m diameter and 2.5 m height. The velocity was measured by sophisticated flow meter, while pressure was monitored by smart transducer. The radial distributions of gas hold-up were investigated under different operating conditions. BCs are characterized by complex fluid dynamic interactions . The hydrodynamics phase mixing, and transports which may affect the performance of BC remains difficult to predict due to the complex phase interactions and flow turbulence.2-D and 3-D distribution profiles of gas hold-up and liquid hold-up were showing conflicted trend between them under similar operating conditions. Stacked distribution profiles were explained the interface between phases. Ug is a major process variable in the BC, which greatly affects the hydrodynamics and flow regimes. Hg increased with Ug, and the liquid holdup Hl has the opposite trend. Gas pressure p is one of important parameter that is affecting the hydrodynamics in BC. An increase in P results in smaller bubbles. As a result, Hg increases because of the longer residence of the smaller bubbles in theBC. Improving of distribution study for BC has required simulating of operating conditions by global mathematical model. The churn-turbulent flow has proven the best flow regime to collect data and formulating of model because of it's characterized of highly-agitated flow and good mixing.The simulated results were obtained with aid of MATlab software tool. Reasonale agreement have obtained when compared the simulated results with the experimental data.

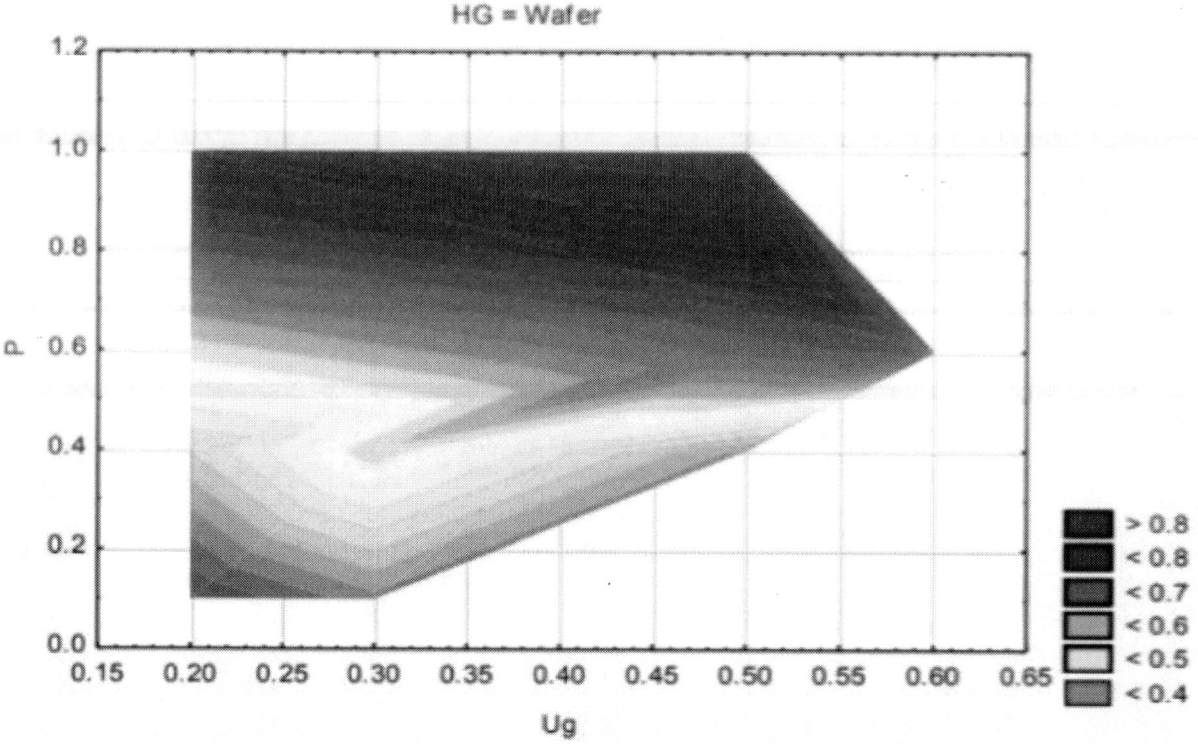
HG = Wafer
1.2
1.0
0.8
0.6
0.4
0.2
0.0
P
0.15
0.20
0.25
0.30
0.35
0.40
0.45
0.50
0.55
0.60
0.65
Ug
> 0.8
< 0.8
< 0.7
< 0.6
< 0.5
< 0.4

Engineering Glycoimmunology

Jessica C. Stark

Immunoengineering promises to yield the next generation of therapies for unmet medical needs in infectious disease, autoimmunity, and cancer. However, a key challenge facing the field is our limited understanding of how the immune system recognizes and responds to glycoconjugates. Glycoconjugates – biopolymers decorated with sugars, or glycans – coat the surface of every cell. As a result, all of molecular immunology is influenced by glycoconjugates. The chemical structures of appended glycans are often altered in disease states and can contribute to pathogenesis by modulating immune responses. Thus, glycoconjugates represent a vast set of attractive, yet mostly untapped, disease-specific antigens and targets for immunotherapy.

My research has resulted in a suite of technologies for elucidating and engineering the immune response to glycoconjugates. I engineered a cell-free, or in vitro, technology for on-demand and portable production of conjugate vaccines that use bacterial cell-surface glycans as antigens. I showed that cell-free synthesized vaccines elicited glycan-specific immune memory that protected mice against lethal pathogen challenge. My cell-free approach can be used for rapid and facile biosynthesis of glycoproteins bearing a variety of user-specified glycan structures, which promises to accelerate interrogation of their immunomodulatory properties. My current work focuses on identifying and targeting glycoconjugates that allow cancer cells to evade anti-tumor immune responses. Specifically, I am developing antibody-lectin bispecifics for blockade of glyco-immune checkpoints in cancer. The bispecific platform is modular and can be applied to diverse disease- or cell type-specific antigens and lectin (glycan-binding) immunoreceptors. In parallel, I developed an interaction proteomics pipeline to define tumor-associated glycoconjugates that engage inhibitory Siglec immunoreceptors. This approach has resulted in the identification of previously unknown immunomodulatory tumor antigens that represent potential targets for cancer immunotherapy. Collectively, my work has helped elucidate the roles of glycoconjugates in disease and promises to accelerate development of new vaccines and immunotherapies.

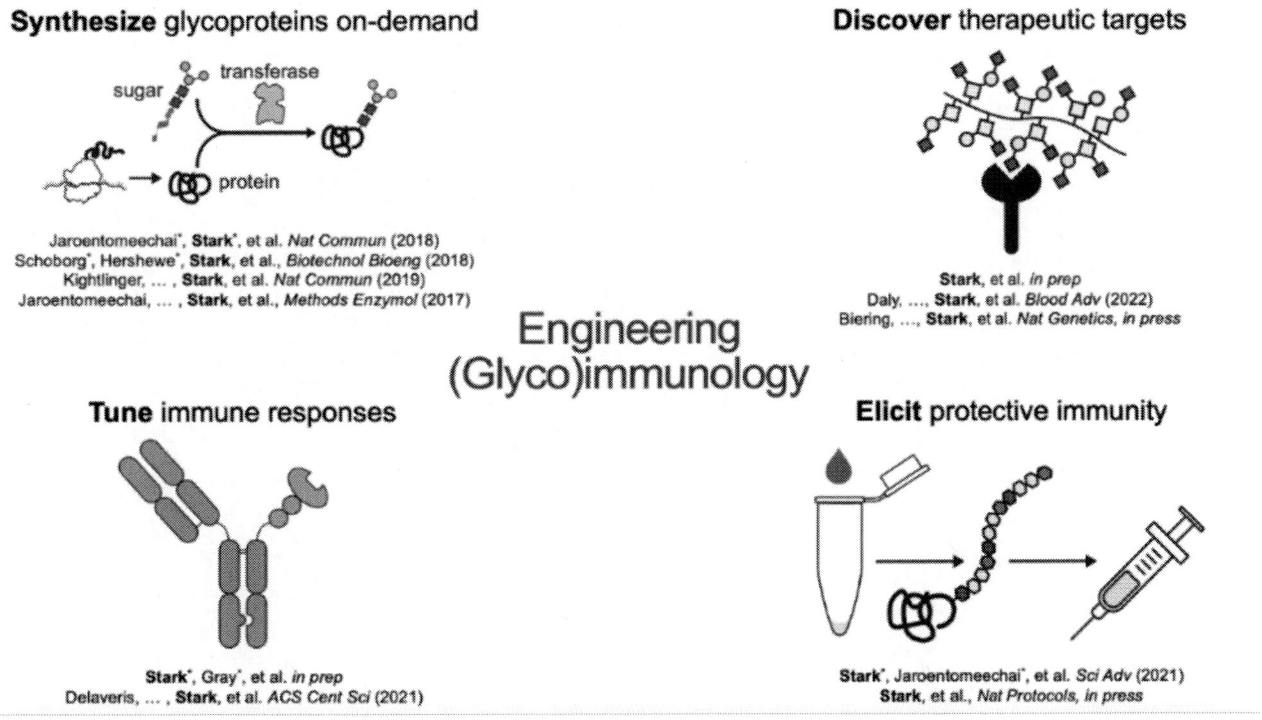

Accurate Prediction of mRNA Degradation at Nucleotide Resolution with Deep Learning

Shujun He, Baizhen Gao, Rushant Sabnis, Qing Sun

Messenger RNA (mRNA)-based vaccines have emerged as rapid and versatile candidates to quickly respond to virus pandemics such as the coronavirus. Pfizer and Moderna have successfully developed effective mRNA vaccines to fight COVID-19. However, one key limitation of mRNA based vaccines and therapeutics is the inherent instability of mRNA molecules. Currently, little is known on where in the backbone of a given RNA is most likely to degrade. Therefore, understanding and predicting mRNA degradation is crucial. In this study, we show how biophysics and deep learning can be combined to learn from experimental data to accurately predict mRNA degradation under different conditions at nucleotide resolution. Our deep learning models achieve up to 0.68 R2 score on mRNA degradation prediction of each nucleotide. Additionally, we demonstrate these predictions are strongly correlated with half-lives of unseen mRNA sequences with up to -0.657 Pearson's r and can thus be used to estimate half-lives of mRNA sequences. Finally, we propose how our deep learning models can be used to stablize mRNA sequences via computational design.

In Situ Electro-Regeneration of Pfas-Laden Adsorbents: Benchmarking Single Use Vs. Reuse

Gamze Ersan, Mahmut S. Ersan, Sergi Garcia-Segura

Perfluoroalkyl substances (PFAS) are anthropogenic chemicals that are persistent and bio-accumulable in the environment. Stringent regulatory compliance measures are being developed at accelerated pace regarding the control of these forever chemical. The only effective proven solution that can be deployed immediately is adsorption technologies due to its removal efficiency, robustness, and low-cost. However, the main concern inherent with PFAS removal through these adsorptive removal techniques is either disposal or regeneration of the exhausted media. To date, there has been no study conducted on the *in situ* electro-regeneration of PFAS-laden carbon-based adsorbents. Therefore, the main objectives of this study are to (i) evaluate electron-driven principles to understand reuse capabilities of adsorbents beyond the breakthrough curve saturation, and (ii) elucidate the impact of PFAS structure and adsorbent characteristics on the electro-regeneration efficiency. In this study, PFAS loaded adsorbents were electrochemically regenerated using carbon (as the cathode) and Titanium wire (as the anode) in a flow-through rapid small scale column test (RSSCT) setup. Based on our preliminary results, the regeneration efficiency of PFOA-laden GAC (F400) could reach up to 40% within 72 hours and no significant change was observed after two consecutive adsorption-regeneration cycle. Additional RSSCT testing is currently underway with mixed PFAS (PFOA [C8], PFOS [C8], PFHxA [C6], PFHxS [C6], PFBA [C4], PFBA [C4]) background, and the results will be available by the end of the July 2022. These experiments will demonstrate the effectiveness of in situ electro-regeneration on mix-PFAS laden GACs.

Effect of Colloids and Ca2+ on Scale Precipitation Rate in RO Systems: A Pilot Plant Study

Nadhem Ismail, Ali Alshami

Inorganic scale formation on RO membrane surface as well as RO concentrate lines threatens the high RO efficiencies and recovery rates. Scale build-up leads to increased maintenance costs, decrease in pipe diameter, and reduce in flow rates. There are different treatment methods to inhibit or remove scale depending on different RO systems. Scale mitigation depends on understanding the factors affecting scale formation such as operating conditions, ions, and presence of colloids. This study concerns the investigation of the effect of Ca2+ and colloids on RO module and RO concentrate lines scaling.

Five closed systems consisting of a pump, PVC piping, clear plastic tubing, and two different plastic tanks (20 L and 50 L) were run under different conditions for several months. The first and second studied the effect of Ca2+ while third, fourth, and fifth systems were designed for colloids testing, short-term, and long-term effects respectively. Scale formation was monitored by means of flowrate, conductivity and pH measurements over time. The findings are as follows: system one, resulted in almost no change in flowrate pointing to minimal scale formation. System two tests resulted in severe case of scale formation that the RO failed after 18 days. System three showed a significant decrease in flow rate after 48 hours of operation. System four showed water flow decrements up to day 35 of operation, then stabilized afterwards. This test also resulted in colloids remaining suspended in solution at various elevated concentrations; but when salts were added, considerable amounts of colloids fell out of solution and some scale formed. System five, in a period of 100 days, revealed a drop in flow by 72 % in the long term piping vs 13% in the short term piping.

Evaluation of the Critical Factors and a Future Towards Machine Learning-Assisted Membrane Fabrication

Musabbir Jahan Talukder, Ali Alshami

Membrane separations offers a green and cost effective replacement of traditional separation processes. For this effective transition, a proper and efficient method for membrane fabrication, that can withstand wide range of process conditions with tunable properties is crucial. Membrane fabrication is a multivariate operation that involves selection and optimization of various parameters. Polymers have been a potential proven material for membrane fabrication. Recently, the large data in polymer membranes in literature is explored and interpreted by machine learning (ML) methods for novel membrane fabrications. With these ML algorithms, an efficient method is devised other than the age-old trials and errors for finding desirable separation performance. In this report, we try to stipulate the recent efforts in terms of critical factors in their investigations. A future direction in material selection and optimization is also outlined in these factors. For a better understanding of the fundamental domain knowledge and experimental experience relation, a novel hybrid modelling structure is discussed for the machine learning assisted membrane fabrication.

Flexible Hydrogen Leak Detectors

Abhishek Ranade, Venugopal Santhanam

Hydrogen is one of the possible replacements for fossil fuels in the future. However, appropriate safety barriers are required due to hydrogen's high flammability and diffusivity. One such barrier is a flexible detector for hydrogen leaks [1]. Commercially available detection tapes can visually detect hydrogen leaks by undergoing a color change [2]. Such tapes require human observation to detect leaks, as there are no accompanying electronics which can continuously read the output. In contrast, chemiresistive hydrogen detectors can continuously read the detector output and signal an alarm accordingly. Palladium is commonly used in the fabrication of chemiresistive sensors due to its high selectivity towards hydrogen. At ambient conditions, hydrogen can be absorbed by palladium to form a hydride, which has a higher resistance than pure palladium. This increase in resistance is indicative of the presence of hydrogen. Thus, a flexible chemiresistive leak detector can signal the presence of hydrogen, without depending on human observation. Detectors having faster responses enable early leak detection, thereby decreasing the risk of accidents. Palladium nanostructures have exhibited very fast response times, and therefore have been widely studied in this regard [3]. Thin polymer sheets, such as polyimide, have excellent dielectric properties and low coefficient of thermal expansion. Large-area polyimide substrates are also commercially available at low-cost, thereby being useful as a substrate material for flexible leak detectors [4]. In this work, electrically conductive Pd nanostructures were deposited on polyimide, using aluminium foil and a Pd salt solution as the sole precursors. This fabrication approach is a bottom-up process performed at ambient conditions [5]. The detectors could respond and recover repeatedly to hydrogen exposures. However, the detector response and recovery became slower with increasing number of exposure cycles (t_{90} ~ 83s @ 3.6% H_2, t_{90} ~ 95 s @ 0.6% H_2). This could be due to the multiple $\alpha \rightarrow \beta$ phase transitions of the palladium hydride creating mechanical stresses in the Pd lattice. Alloying of Pd with other metals is one strategy to mitigate this phase transition e.g., Pd-Pt nanostructures have been shown to exhibit faster responses in comparison to Pd [6]. Therefore, bimetallic Pd-Pt nanostructures were deposited on polyimide by a simple modification of the precursor composition. The detectors fabricated using 95 mol% Pd were found to be faster than the Pd-based detector. e.g. after undergoing the same cycling test as the Pd-based detector, the bimetallic Pd-Pt detector exhibited faster responses (t_{90} ~ 40s @ 3.6% H_2, t_{90} ~ 79 s @ 0.6% H_2).. Thus, rapid and flexible chemiresistive detectors could be fabricated using a low-cost, facile approach that is amenable for scale-up using roll to roll techniques. In this talk, I will describe the process developed for sensor fabrication and analyse the results of hydrogen leak detection tests.

References:

[1]: Hübert, T., Boon-Brett, L., andamp; Buttner, W.. *Sensors for safety and process control in hydrogen technologies.* Florida, FL: CRC Press; (2016): pp 16-17.

[2]: Roberson, L., Captain, J., Williams, M., et al. Chemochromic Hydrogen Leak Detectors. NASA Tech Briefs. 2009; 33 (6), pp 54.

[3]: Koo, W. T., Cho, H. J., Kim, D. H., Kim, Y. H., Shin, H., Penner, R. M., andamp; Kim, I. D. Chemiresistive hydrogen sensors: fundamentals, recent advances, and challenges. *ACS nano.* 2020; 14(11), pp 14284-14322.

[4]:Rashid, T. R., Phan, D. T., andamp; Chung, G. S.. A flexible hydrogen sensor based on Pd nanoparticles decorated ZnO nanorods grown on polyimide tape. *Sensors and Actuators B: Chemical.* 2013; *185*, pp 777-784.

[5]: Li, W., Cochell, T., andamp; Manthiram, A.. Activation of aluminum as an effective reducing agent by pitting corrosion for wet-chemical synthesis. *Scientific reports.* 2013; *3(1), pp 1-7.*

[6]: Li, X., Liu, Y., Hemminger, J. C., andamp; Penner, R. M.. Catalytically activated palladium@ platinum nanowires for accelerated hydrogen gas detection. *ACS nano.* 2015; *9(3), pp 3215-3225.*

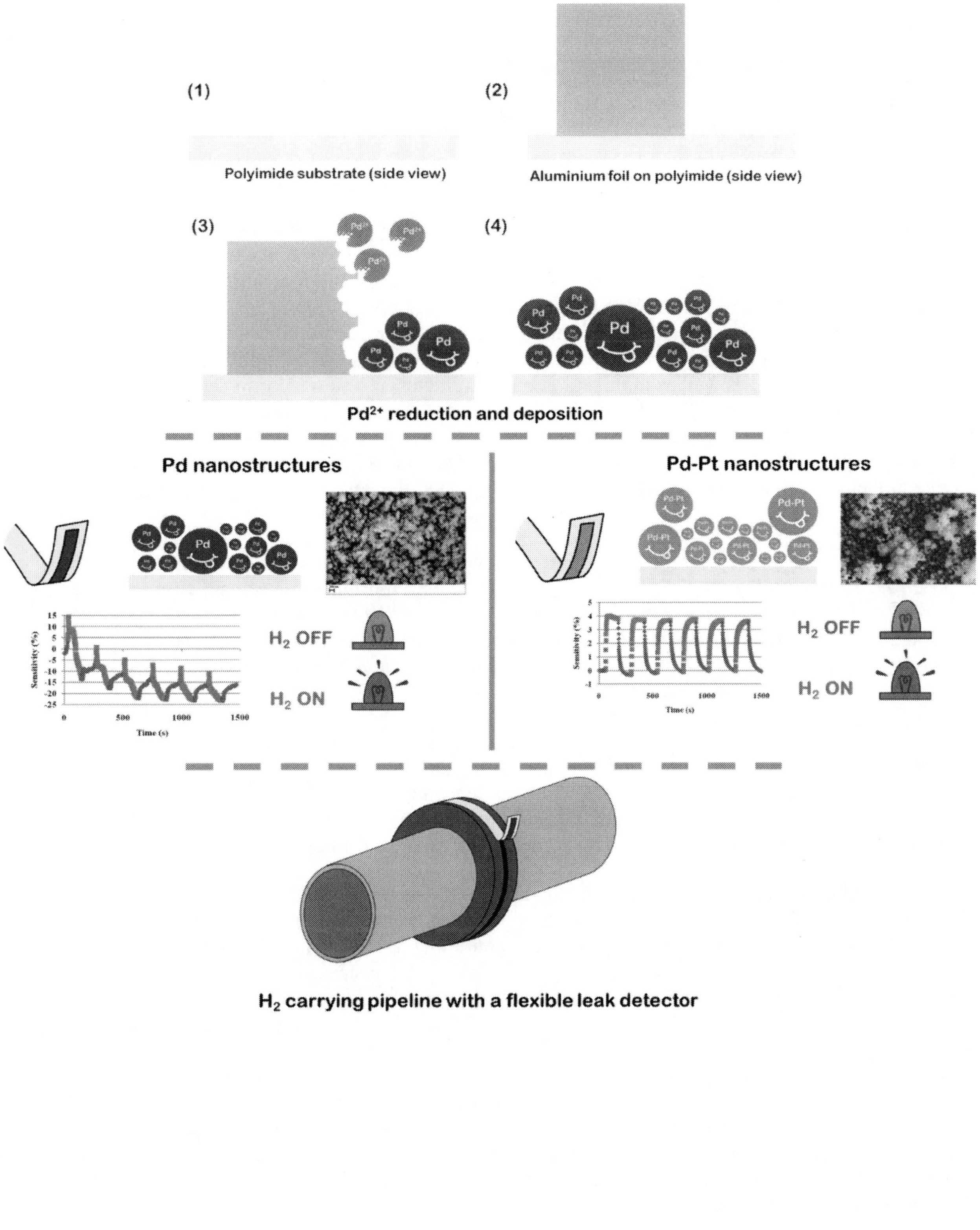

A Semantic Based Decision Support Framework to Enable Model and Data Integration

Linsey Koo, Edlira Kalemi

Process modelling and simulation are employed as vital tools to design and evaluate biorefining technologies. With an increasing number of models and data built by various computer aided process engineering (CAPE) tools in the domain of biorefining, the researchers who stand to benefit from such an updated understanding are manifold. However, this large number of models and data distributed from CAPE tools results in heterogeneity and further requires expertise to establish interoperability.

The emergence of developing a common interface has led to the CAPE-OPEN initiative. CAPE-OPEN is a standard that defines rules and interfaces that allows CAPE tools and/or models to interoperate at a software engineering level. The CAPE-OPEN mainly focuses on the interoperability of flowsheet simulation. The interoperability is established between the process modelling components (PMCs) that are developed in various process modelling environments (PME) using a "middleware" The middleware is a connectivity mechanism that hosts communication across heterogeneous modelling environments by providing their functions, as well as input and output parameters (CO-LaN consortium, 2003). Despite the awareness and visibility of CAPE-OPEN within chemical engineering community, most end users without much software engineering expertise provided feedback that some found it difficult to use as it requires expert level support (Baur, 2018). To gain a fuller ability to increase awareness and to retain the most suitable model and/or data amongst existing models and data, the development of a systematic approach is required.

In this paper, a new approach for model integration however builds upon the CAPE-OPEN framework proposes the use of ontology and replaces the middleware with more flexible semantic repository ((Koo, Trokanas and Cecelja, 2017). Models are described by Semantic Web Services (SWS) using Ontology Web Language for Services (OWL-S) as an enabler of web services through service discovery, selection, composition, and execution stages (Figure 1). By The same token, datasets are also described by the OWL-S through their outputs and auxiliary precondition for their execution, i.e. extraction, modification and deletion (Figure 2). The process of model and data discovery using input/output matching has been supported by the structure of domain ontology. The matching process was employed as a method to perform the flexible discovery of models and data, which allows partial matching for inputs and outputs. Users are then given choices of ranked discovered models and data based on the requirements to assist them to make an inform decision. Once the relevant models and data sets are chosen, the demonstration of establishing interoperability is performed by composition and execution process.

This paper mainly focuses on the key parameters used in the formation of model integration through input/output matching based on semantic relevance between the models and data. The heterogeneous models and data in the domain of biorefining are semantically described in a domain ontology, OntoCAPEmodel ontology. The ontology acts as a framework that serves as a blueprint for knowledge dissemination and exchange by defining the basic terminology and interaction between models and data. The structure of unit is configured by a coupling through the different inlet and outlet ports where a unit can be connected to another unit, which is separated from the functional behaviour of the unit model. A set of data to be exchanged from one unit to another is distinguished by three different types: material, energy, and information streams. In addition, the stream properties are characterised by thermodynamics and physical properties, e.g. composition, temperature, pressure, flow, etc., with associated variables describing quantitative properties, e.g. physical dimension, value, unit, etc. The structure of the ontology further guides a process to record details about the models and data in an explicit format, and concomitantly supports registration and instantiation of the models and data.

The whole process is verified by a case of model integration for a supply chain modelling in biorefining.

References

1. Baur, R. (2018) CAPE-OPEN Survey CAPE-OPEN Survey - Objectives, CAPE-OPEN 2018 Annual Meeting, CO-LaN. Available at: http://www.colan.org/presentation/cape-open-survey-2/.
2. CO-LaN Consortium (2003) CAPE-OPEN Methods andamp; Tools Integrated Guidelines, Cape-Open.
3. Koo, L., N. Trokanas, et al. (2017). 'A semantic framework for enabling model integration for biorefining.' Computers andamp; Chemical Engineering 100: 219-231

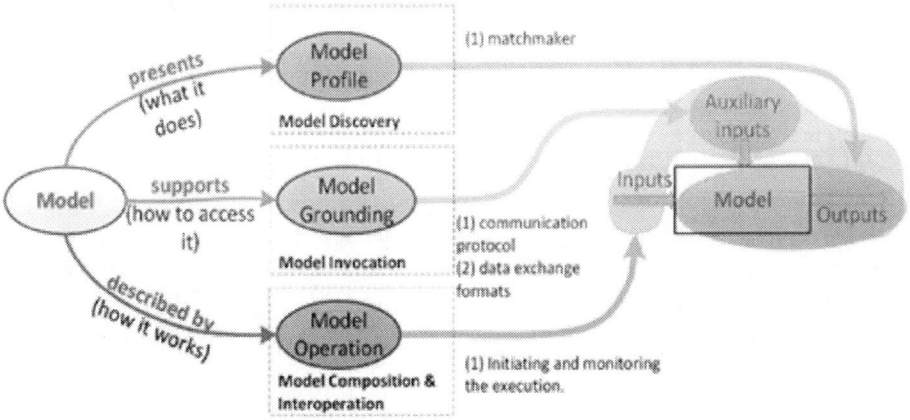

Figure 1 Semantically described model

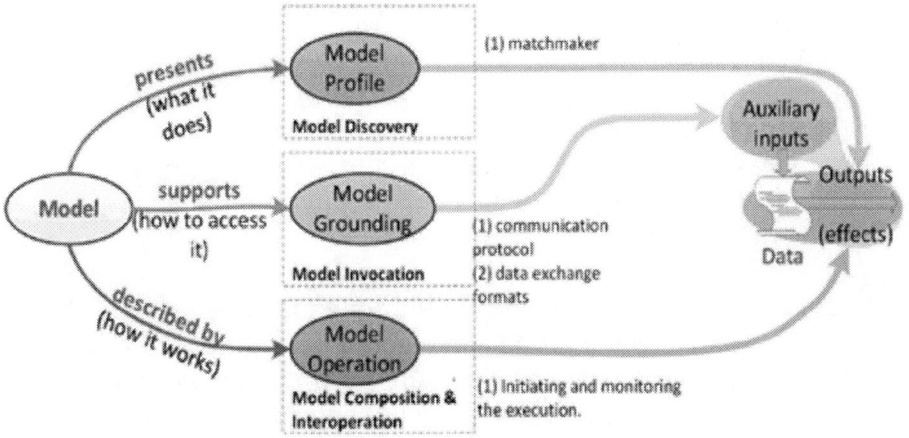

Figure 2 Semantically described datasets

Product and Host Engineering for Low-Cost Manufacturing of Therapeutic Proteins in Yeast

Neil C. Dalvie

I recently completed my PhD in Chemical Engineering at MIT, advised by Professor J. Christopher Love, and am seeking an independent research and teaching career. I am interested in industrial applications of microorganisms as renewable cellular catalysts. In this talk, I will describe my graduate research on manufacturing of therapeutic proteins, and briefly describe my future research goals.

The COVID-19 pandemic revealed a global need for affordable, accessible biologic medicines such as prophylactic vaccines and antiviral monoclonal antibodies. Indeed, two years after the emergence of SARS-CoV-2, access to vaccines and therapeutic proteins is still limited in low- and middle-income countries (LMICs). Long development timelines, expensive manufacturing requirements, and stringent storage conditions all contribute to the high cost of biologic medicines generally. Traditionally, complex therapeutic proteins are produced in mammalian cells, which increase costs due to sterility requirements and expensive feedstocks, and extend development timelines due to slow cell growth.

Alternative hosts have potential to increase the accessibility of therapeutic proteins by shortening development timelines, lowering manufacturing costs, and increasing the global manufacturing capacity. Single-celled eukaryotes such as yeasts represent a "sweet spot" for protein manufacturing —yeasts grow to high cell densities on inexpensive feedstocks like bacteria, and are capable of post-translational modifications and secretion of products into the extracellular space like mammalian cells. One yeast, *Komagataella phaffii* (*Pichia pastoris*), is especially attractive due to a highly developed secretory pathway. *K. phaffii* is currently used for manufacturing of insulin and subunit vaccines in LMICs, and has recently been approved by the FDA for the manufacture of several complex therapeutic proteins, including a monoclonal antibody. Wider adoption of alternative hosts such as *K. phaffii* in both high-income countries and LMICs has been hampered by bespoke manufacturing challenges for unique therapeutic proteins such as subunit vaccines, and low upstream titers of complex therapeutic proteins with stringent quality requirements such as monoclonal antibodies.

In my PhD research, I explored two strategies to improve manufacturing of therapeutic proteins– protein engineering of the therapeutic product and genomic engineering of manufacturing cell lines. In this talk, I will describe two illustrative case studies of integrated product and manufacturing process design. First, I will describe design and manufacturing of a subunit vaccine candidate for COVID-19. Engineering of the protein antigen improved the manufacturability and potency of the vaccine, and engineered strains enabled rapid technology transfer and scale up. This candidate is currently in clinical trials. Second, I will describe a platform for reliable manufacturing of monoclonal antibodies in *K. phaffii*. I engineered a yeast strain with improved productivity, and an IgG1 backbone sequence with improved product quality. We demonstrate that this yeast-based platform has comparable quality and productivity to typical manufacturing processes in mammalian cells.

These successes demonstrate the potential of microorganisms to disrupt and improve existing industrial processes. In the case of therapeutic proteins, the ability of the yeast *K. phaffii* to secrete the product into the extracellular space enables rapid, low-cost manufacturing, because cells act as a renewable catalyst. I will end my talk with a brief description of my future research aims to use microorganisms as catalysts to mediate the build-up or breakdown of complex products and materials to address a range of global problems.

Molecular Subgraphs in AMS2022: Applications in Molecular Dynamics, Molecule Clustering, Property Prediction, and the Inverse Design Problem

Nick Austin, Fedor Goumans

Molecular subgraphs, sometimes referred to as signature descriptors[1], constitute a useful set of descriptors that are applicable in many domains of cheminformatics (e.g., Morgan fingerprints[2]) and computational chemistry. Molecular subgraphs contain information about a central atom and its local environment, typically to within a certain radius of bonds away from the central atom. In this talk, we will discuss the development of large databases (500,000+ entries) of subgraphs, the many possible features (e.g., bond orders, stereochemistry, depth, implicit hydrogens, etc.) we choose to use for subgraphs, and the effect of including/removing these different features has on the representative power of subgraph descriptors.

Additionally, we will discuss applications in the context of our computational chemistry software, AMS2022[3]. First, we will discuss subgraphs as applied to the analysis of reactive molecular dynamics (MD) trajectories. Next, we will describe the usage of subgraphs to cluster similar molecules both within MD trajectories and in more general cases. Subgraphs can additionally be applied to the prediction of certain properties. For property prediction, each unique subgraph is associated with a computed or regressed vector of descriptors, and the property of the molecule is given by a (typically simple) function of these descriptors for each constituent subgraph. This approach is compared to current machine-learning-based approaches in terms of accuracy and prediction domain. Finally, we present a mixed-integer linear programming (MILP) formulation for the inverse design problem[4] based on subgraphs and apply this to an example molecular design problem.

1. Faulon, Jean-Loup, Donald P. Visco, and Ramdas S. Pophale. 'The signature molecular descriptor. 1. Using extended valence sequences in QSAR and QSPR studies.' *Journal of chemical information and computer sciences* 43.3 (2003): 707-720.
2. Rogers, David, and Mathew Hahn. 'Extended-connectivity fingerprints.' *Journal of chemical information and modeling* 50.5 (2010): 742-754.
3. ADF2022, SCM, Theoretical Chemistry, Vrije Universiteit, Amsterdam, The Netherlands, http://www.scm.com.
4. Austin ND, Sahinidis NV, Trahan DW. Computer-aided molecular design: an introduction and review of tools, applications, and solution techniques. Chemical Engineering Research and Design. 2016 Dec 1;116:2-6.

Treatment of Cooling Water Blowdown (CWBD) Using Vibratory Sheared Enhanced Process (VSEP)

Mariam Soliman, Fadwa Eljack, Kazi Khoda, Fares Almomani, Elalim Ahmed, Ziad Eljack

Cooling water blowdown (CWBD) is a concentrated wastewater effluent from cooling towers that are used in different applications including cooling systems[1]–[3]. The large-scale cooling needs is supplied in the State of Qatar by district cooling facilities to meet the high demand of various industries including university campuses, commercial communities, business districts, etc. CWBD has high concentrations of various contaminants, which deem them unsuitable for recycling, reusing or discharging into wastewater treatment plants and surface water. Using an end of pipe treatment process is considered to enhance the effluent water quality to deem it suitable for recycling within cooling systems, or for discharge to surface water without considerable negative consequences. Several treatment technologies are available for the treatment of wastewater such as electrocoagulation (EC), electrodialysis (ED), nanofiltration (NF), and reverse osmosis (RO). Some of these technologies have high sludge discharge such as ED, others such RO have a fouling issues that can affect the performance of the process[4]. In this work, we investigate the implementation of vibratory sheared enhanced process (VSEP) as an emerging membrane technology for the treatment of CWBD at lab and pilot scale. VSEP offer the advantage of using a mechanically vibrated membrane over conventional membrane technologies that reduce the fouling problem associated with the membrane and give a better TDS removal efficiency, as high as 96.5%[5]. The enhanced performance is due to the high sheared stress applied on membrane surface [6, 7]. At the lab scale, nanofiltration, microfiltration, and reverse osmosis membranes will be used in membrane selection study to decide upon the most suitable one based on major parameters such as the permeate rate stability and its quality. Additionally, a pressure study will be performed to choose the optimum pressure and use it for the concentration study to know the recovery percentage. Key parameter such as pressure, temperature, flowrate, pump speed, and vibration speed will be analyzed to understand their effect on the permeate stability as well as its quality. The quality of CWBD is tested by measuring the total dissolved solids (TDS), conductivity, as well as the ions concentrations using ionic chromatography before and after the treatment. This study will be conducted at both the bench scale and and the system will be scale up to pilot. The findings of the study will provide insights into the potential of using VSEP in cooling facilities and related municipalities a treatment alternative for the management of CWBD.

References

[1] H. I. Abdel-shafy, M. A. Shoeib, M. A. El-khateeb, A. O. Youssef, and O. M. Hafez, "Electrochemical treatment of industrial cooling tower blowdown water using magnesium-rod electrode," *Water Resour. Ind.*, vol. 23, p. 100121, 2020.

[2] O. M. Hafez, M. A. Shoeib, M. A. El-Khateeb, H. I. Abdel-Shafy, and A. O. Youssef, "Removal of scale forming species from cooling tower blowdown water by electrocoagulation using different electrodes," *Chem. Eng. Res. Des.*, vol. 136, pp. 347–357, 2018.

[3] E. A. Anwer and B. A. Abdul Majeed, "Different Electrodes Connections in Electrocoagulation of Synthetic Blow down Water of Cooling Tower," *Iraqi J. Chem. Pet. Eng.*, vol. 21, no. 1, pp. 1–7, 2020.

[4] M. Soliman, F. Eljack, M. K. Kazi, F. Almomani, E. Ahmed, and Z. El Jack, "Treatment

technologies for cooling water blowdown: A critical review," *Sustain.*, vol. 14, no. 1, 2022.

[5] G. Johnson, L. Stowell, and M. Monroe, "A comparison of Conventional Treatment methods and VSEp, a Vribrating Membrane Filtration System," *Desalin. Confèrence*, 2006.

[6] J. Leong, J. Tan, A. Heitz, and B. P. Ladewig, "Use of vibratory shear enhanced processing to treat magnetic ion exchange concentrate: A techno-economic analysis," *Desalination*, vol. 383, pp. 46–52, 2016.

[7] A. I. Zouboulis, E. N. Peleka, and A. Ntolia, "Treatment of tannery wastewater with vibratory shear-enhanced processing membrane filtration," *Separations*, vol. 6, no. 2, 2019.

Hypersaline Brine Desalination Using Solvent-Driven Water Extraction with Dimethyl Ether

Akshay Deshmukh, Zi Hao Foo, Aaron D. Wilson, John H. Lienhard

Solvent-driven water extraction (SDWE) has promising applications in high-salinity brine desalination, including zero-liquid discharge processing for industrial wastewaters, and resource recovery, such as the extraction of lithium and rare earth elements from solution-mining leachate. In this study, we develop a novel SDWE system combining a liquid-liquid separator (LLS) with a three-phase mechanical vapor compression (3P-MVC) process to efficiently desalinate hypersaline brines using dimethyl ether (DME). DME is a polar aprotic solvent that is partially miscible with water. The low polarity of DME minimizes the solubility of electrolytes, such as sodium chloride (NaCl). The 3P-MVC system enables the rapid separation of water-DME mixtures after extraction, incorporating effective heat recovery to minimize energy consumption.

We first describe a computational platform to calculate the excess Gibbs free energy of water-DME-NaCl mixtures using the electrolyte non-random two-liquid (eNRTL) model. Interaction parameters are estimated from fluid-phase equilibrium and osmotic coefficient data using metaheuristic global optimization techniques for NaCl concentrations from 0.0 to 6.0 mol kg^{-1}. A new system-scale computational model combining a counter-current, multistage LLS with 3P-MVC is then applied to explore the water recovery and specific energy consumption of SDWE as a function of solution composition, system temperature, and operating pressure. For a hypersaline brine with a salinity of 2.0 mol kg^{-1} (more than three times seawater salinity), our analysis demonstrates that SDWE using DME can achieve a water recovery ratio greater than 50% and a specific energy consumption under 60 $kWh_e\ m^{-3}$. We conclude by exploring the wider performance envelope of SDWE and performing a system-scale optimization to maximize water recovery and energy efficiency across a range of feed salinities.

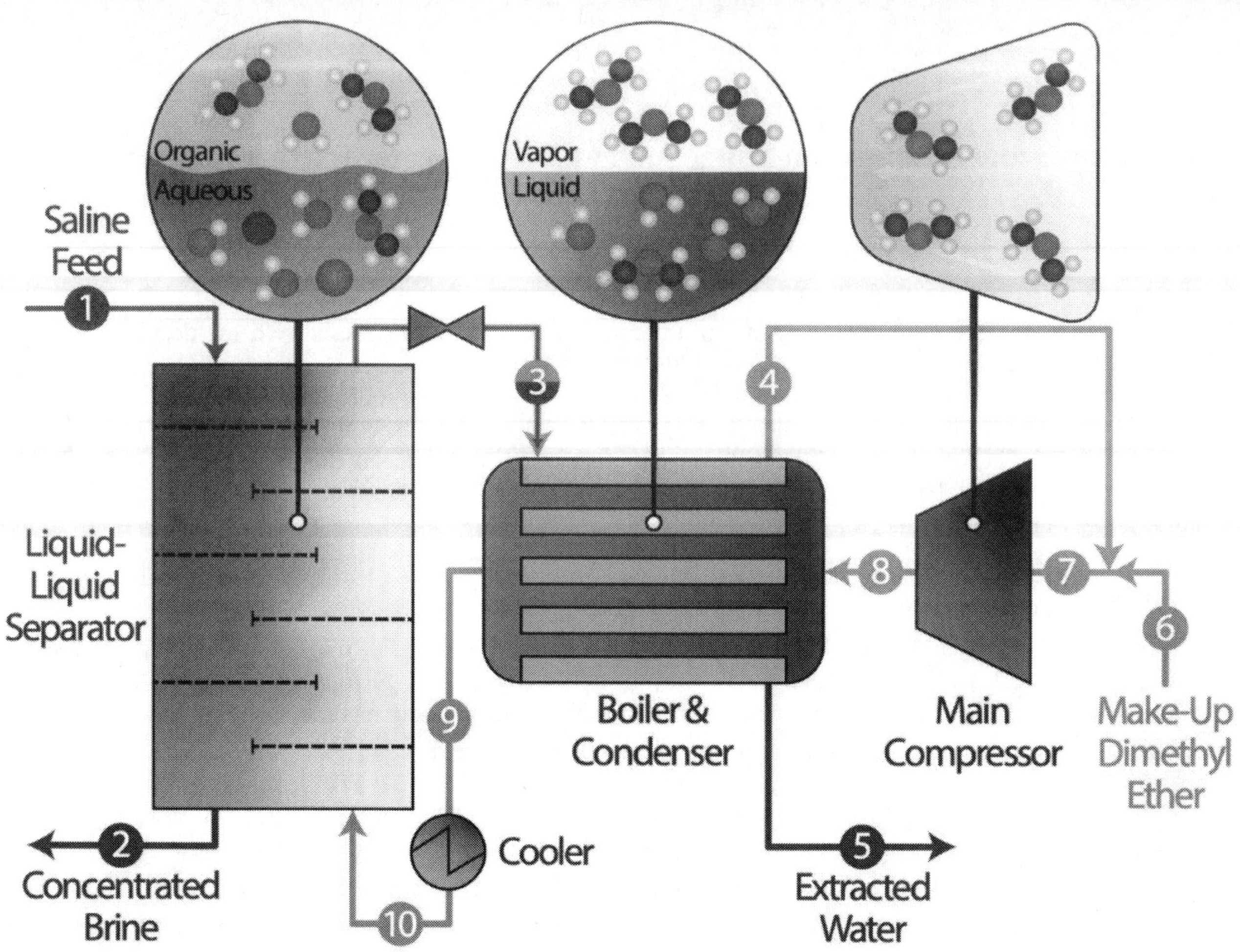
Organic
Aqueous
Vapor
Liquid
Saline Feed
1
3
4
Liquid-Liquid Separator
8
7
6
9
Boiler & Condenser
Main Compressor
Make-Up Dimethyl Ether
2
Concentrated Brine
Cooler
10
5
Extracted Water

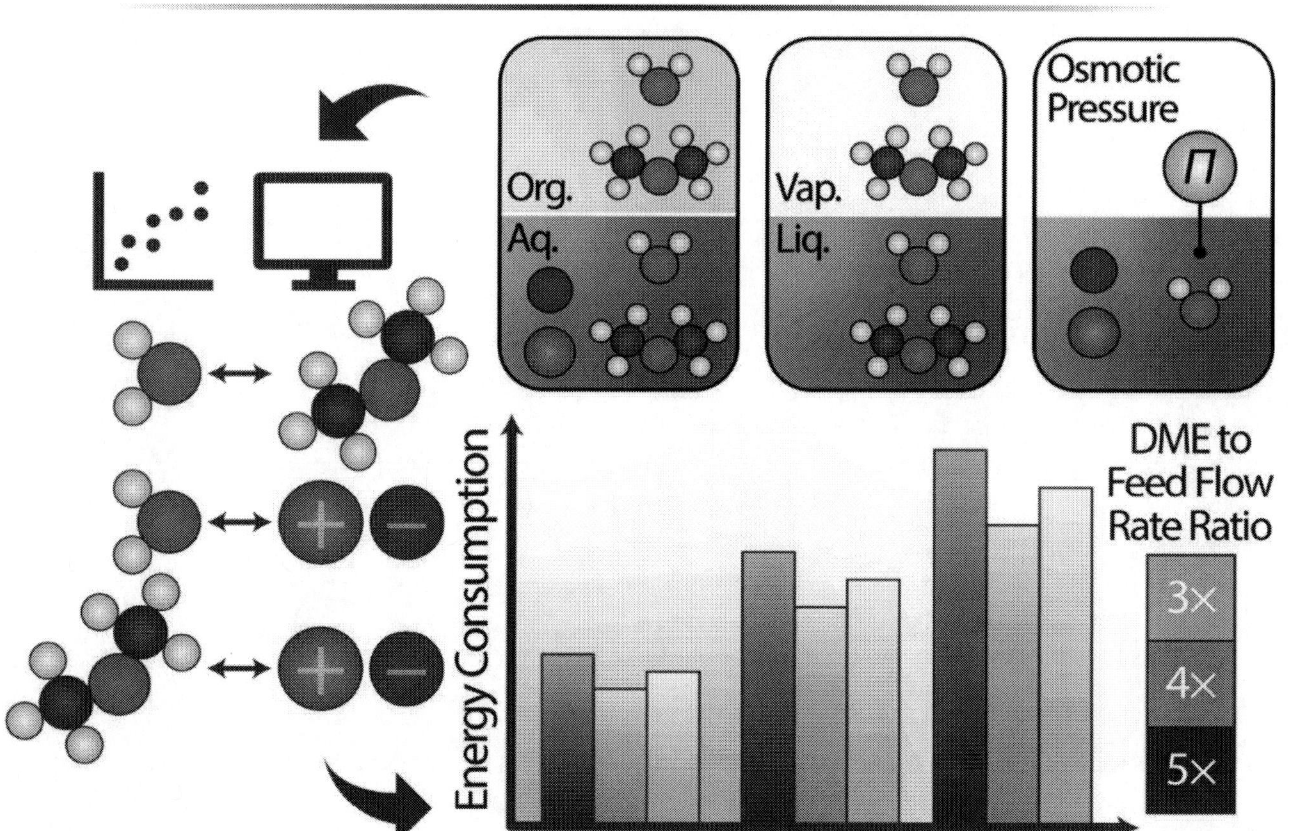
Org.
Aq.
Vap.
Liq.
Osmotic Pressure
Π
Energy Consumption
Hypersaline Brine Salinity
DME to Feed Flow Rate Ratio
3×
4×
5×

Running Machine Learning Learning Models in Chemical Operations

Sebastian Werner

Modeling of chemical processes has been a vital tool for engineers in the chemical industry for decades, and incorporation of such models into daily operations is also not uncommon.

This includes classical first-principles models (flowsheet models) or quite recently data-driven models, as well as so-called hybrid models. Applications of models include, but are not limited to, design of reactors and entire plants, operations planning, scenario analysis or optimizations such as debottlenecking. All of the aforementioned are typically off-line, meaning that they are not done with live data and directly in operations, which is often attributed to the fact that models are steady-state rather than dynamic.

Whereas first-principle models are often built in proprietary expert software making use of parameters from databases like DIPPR, data-driven models rely primarily on observational (experimental) data. Often, first-principle models include fit parameters where certain parameters of correlations are fit to experimental data in a batch process using techniques such as least squares or maximum likelihood. As we are working with continuous operations, the data in focus is typically time-series data collected by sensors in the process control system and stored and analyzed in historians.

Taking this approach even further there are purely data-driven models which use arbitrary relationships to which the observations are fit to. Model structures in this approach can be neural networks or advanced deep learning systems. However also here, parametrizing or typically called "training" such models is often a tedious, time consuming process which is also done with off-line data.

For continuous operations we employ process control systems based on many PID control loops - fed by the current sensor signals. This of course raises the question of the optimal set-points for these individual control loops: Determining those can be based on experience / best-practice handbooks / operational manuals or on the output of an off-line model of the process. Additionally several methods of so-called supervisory control systems have been successfully employed, e.g. advanced process control systems. As rigid first principles models bear the advantages of stability, in an ever changing process the adaptation to reality remains challenging. Contrasting to that, the ability to adapt and learn is an often cited advantage to neural networks and related techniques. It has been a focus of research, not just in chemical engineering and process control, to bring these two approaches together - especially with safety in mind.

Based on recent reports by several consulting firms over 80% of all data-driven proof-of-concepts never make it into a production setup (VentureBeat 2019, Gartner 2019). This is particularly disturbing as this means that these projects have produced significant cost, but never realized the benefits. Very much related to this is the observation that for every $1 you spend on developing a machine learning algorithm, you will spend $100 to deploy and support it. In this contribution, we want to specifically focus on the integration of data-driven models into continuous operations - and the pitfalls and best-practices, with an emphasis on the learnings from operating data-driven models from other disciplines.

With the example is the approximate forecasting of electricity demand and CO_2 emissions forecasting based on a time series of electricity demand consumptions we will outline the whole lifecycle around such an integration. Whilst being rather trivial, it illustrates the challenges in the integration of OPC-UA sources and benefits of running a simple data driven model to gather such information. The attached figure shows an examplary output of such a time-series model. We will outline the whole development cycle from integration, data exploration, cleansing into validation and verification. Subsequently we outline the productionalization of said model as well as the continuous monitoring to maintain model stability as well as keep an eye on well-known problems such as model and data drift.

Summarizing this contribution focuses on giving an overview on the best practices based on numerous projects realized in the process industries that successfully bridge the gap.

[1] Dynamic Modeling of Chemical Reaction Systems with Neural Networks and Hybrid Models in Chemical Engineering andamp; Technology by Hans-Jörg Zander, Roland Dittmeyer, Josef Wagenhuber (8 July 1999) https://doi.org/10.1002/(SICI)1521-4125(199907)22:7%3C571::AID-CEAT571%3E3.0.CO;2-5

[2] Solle, Dörte, et al. 'Between the poles of data-driven and mechanistic modeling for process operation.' Chemie Ingenieur Technik 89.5 (2017): 542-561.

[3] Dobbelaere, Maarten R., et al. 'Machine learning in chemical engineering: strengths, weaknesses, opportunities, and threats.' *Engineering* 7.9 (2021): 1201-1211.

[4] Schweidtmann, Artur M., et al. 'Machine learning in chemical engineering: A perspective.' Chemie Ingenieur Technik 93.12 (2021): 2029-2039.

[5] Russell, Stuart (October 8, 2019). Human Compatible: Artificial Intelligence and the Problem of Control. United States: Viking. ISBN 978-0-525-55861-3. OCLC 1083694322.

[6] https://venturebeat.com/2019/07/19/why-do-87-of-data-science-projects-never-make-it-into-production/

[7] https://blogs.gartner.com/andrew_white/2019/01/03/our-top-data-and-analytics-predicts-for-2019/

[8] *Getting Serious About Data and Data Science,* Thomas Redman and Tom Davenport, Sloan Management Review - September 28, 2020

[9] https://blog.dataiku.com/electricity-consumption-co2-emissions-solution

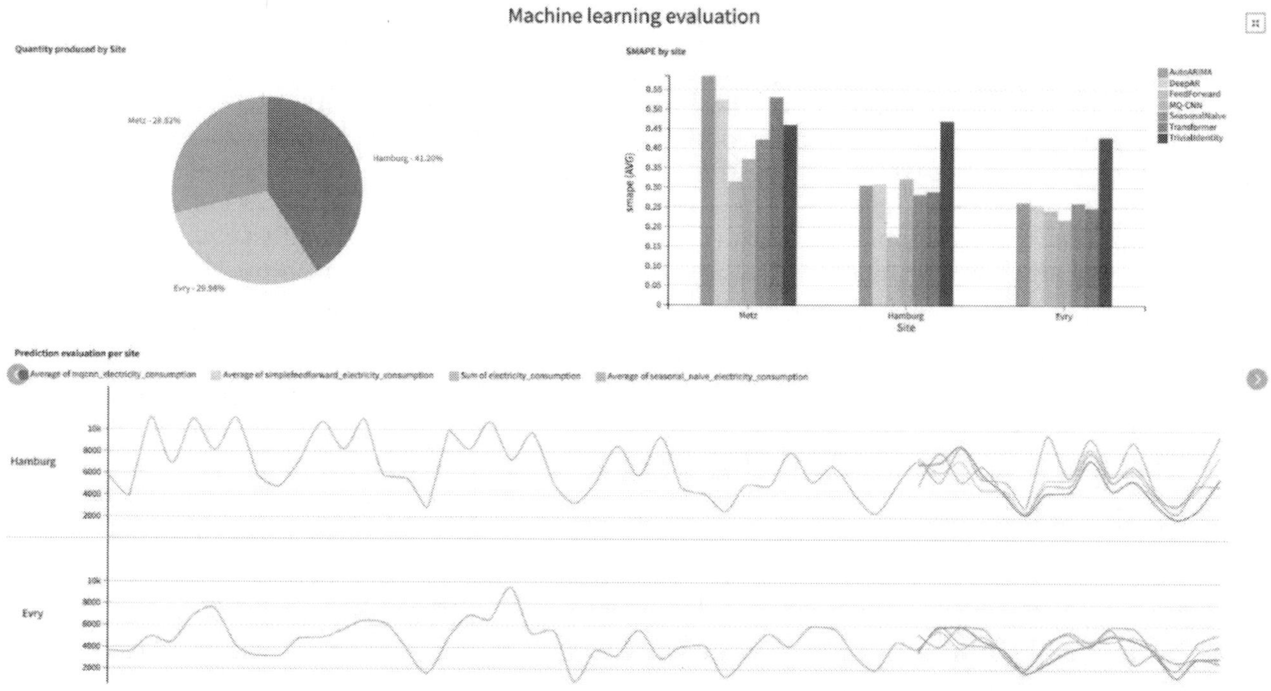

Monitoring Nitrate and Salt Passage, Their Correlation, and Data Imputation Via Data-Driven Process Models for Distributed Membrane-Based Water Purification Systems

Nora Marki, Yang Zhou, Christian Aguilar, Bilal Khan, Yoram Cohen

Groundwater supplies in agricultural regions are often impaired due to high levels of nitrate and salinity. As a consequence, such water supplies need to be treated for both nitrate removal and salinity reduction in order to render these water supplies suitable for potable use. In particular, it is noted that nitrate is a contaminant that can lead to methemoglobinemia ("blue baby syndrome") and increased risk of recurrent respiratory infections, thyroid dysfunction, and certain cancers. Purification of such impaired water supplies has been shown to be effective via integrated filtration and reverse osmosis (RO) membrane systems. However, there are strict regulatory requirements mandate real-time monitoring of nitrate and salinity levels in the product water. The cost of online spectroscopic nitrate sensors is high and can comprise up to about 25% of the total treatment system capital cost, while conductivity sensors (for salinity monitoring) cost is 10% of the nitrate sensor cost. Therefore, means of assessing membrane performance with respect to nitrate based on monitoring of salt passage is critical and can potentially reduce the requirement for online nitrate sensors and thus lower the treatment system capital cost. Accordingly, in the present study, the correlation of nitrate and salinity passage was investigated for a wellhead water treatment system operating in a small disadvantage community in Salinas Valley, California. Data acquired over a period of ten months was processed , and whereby the nitrate sensor was periodically calibrated under laboratory conditions and in the field based on laboratory analysis of product water grab samples. Correlation of the nitrate and salt passage data indicated a high degree of correlation suggesting that nitrate passage can be used as a surrogate for nitrate passage (accounting for data normalization with respect to the feed quality). High resolution data (collection frequency of 1Khz) was utilized showing linear correlation (with a coefficient of determination in the range of 0.91-0.97 between nitrate and salinity of the permeate product water in the different modes of system operation (startup, steady-state production, and shut down). In addition, derived ANN process models were then utilized to describe system operation during the above operational modes under a wide range of operating conditions. Data imputation with this models was shown to be highly reliable and along with the confirmation of nitrate-conductivity correlation one can assert that with the present approach water system operators should be able to utilize salt passage 9via conductivity measurements) as a surrogate for quantifying nitrate levels in the permeate product water; hence, reducing reliance on presently costly online process nitrate sensors and thus reducing the cost of deploying distributed water treatment systems in small and remote disadvantaged communities.

Improvement of Electrodialysis Performance By Surface Modification of Cation Exchange Membranes with Chitosan

Matthew Sheorn, Santanu Kundu

Electrodialysis is a technology that can be used for various separations including desalination. This process is very efficient but can suffer from issues at low ionic concentrations. By modifying the membrane surface of a commercially available cation exchange membrane (CEM), this study aims to show how increasing hydrophilicity improves separation performance. Chitin was deacetylated using a heated NaOH solution over various time intervals. When deposited on the membrane surface, Chitosan changed the contact angle of the CEM. Contact angles were decreased by 21%, 47%, and 62% from the control when coated with Chitosan with degrees of deacetylation of 15%, 30%, and 50%, respectively. Coating success was determined using IR techniques, contact angle measurement, and SEM. A base membrane as well as membranes coated with Chitosan of varying degrees of deacetylation were tested to determine how common electrodialysis parameters, such as limiting current density (LCD), ionic flux and current efficiency are affected by the level of membrane hydrophobicity. Special attention was paid to performance improvements in low ionic concentration areas (less than 1000 ppm NaCl).

Engineering Biological Systems for Human Health and Climate Resilience: From Proteins to Ecological Communities

Tejas Navaratna

Microbes in ecologically diverse communities have evolved complex survival strategies, including producing bioactive secondary metabolites. Compounds from fungi and other species have been exploited for centuries by humans for use as medicines, most notably in the anti-cancer and antibiotics space. These naturally-derived products have been credited with greatly reducing mortality globally. However, there is broad medical consensus that we are rapidly approaching an antibiotic "cliff", where microbial drug resistance will render most antibiotic treatments useless. Recently identified pan-drug resistant *Candida* and *Cryptococcus* fungi are a looming threat especially in healthcare settings. A major focus of my lab will be the high-throughput discovery of peptide and small molecule antifungals and antibacterials. My group will also leverage the power of non-canonical amino acid-based proteomics to gain insight into the mechanisms of persistence and resistance in multi-drug resistant fungi and bacteria.

Furthermore, our group will be tackling the growing climate crisis and it is clear that we need to move away from petroleum-based products including fuels, plastics, and other raw materials. Plants and green algae remove billions of tons of carbon from the atmosphere and convert it to cellulose and other biopolymers. Fungi and other microorganisms break down this biomass, transforming it into metabolically useful products. The second major component of my research program will focus on engineering diverse fungi, including white rot fungi capable of lignin degradation and anaerobic gut fungi found within herbivore rumens, within a community of supporting microorganisms for producing valuable bio-based products such as medium and long-chain fatty acids, bioplastics, and biofuels. I will apply principles of synthetic auxotrophy and cooperation to generate customizable and productive communities of microorganisms.

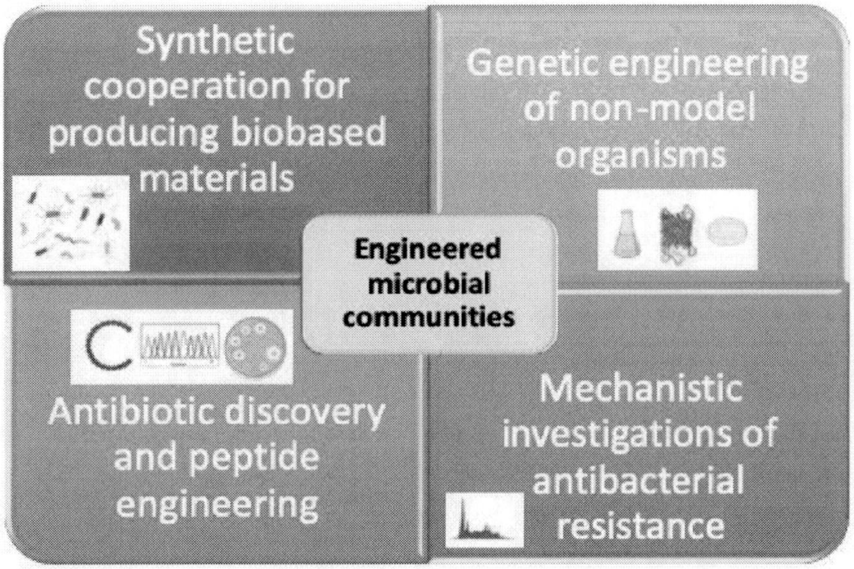

Developing Reusable and Scalable Alternative Filtration Membrane Using Centrifugal Spinning for Personal Protective Equipment

Dhanya Venkataraman, Jay Park

The recent COVID-19 pandemic had impacted our society and, accordingly, the use of personal protective equipment (PPE) has been critical to physically mitigate the spread. The use of meltblown electrostatic-charged polypropylene, while effective, has put strain on supply chain as well as the increasing plastic/fabric waste issues. A strong need to create an economical, reusable, with comparable or improved filtration is highly desired. In this work, we utilize centrifugal spinning to develop filter layer with comparable diameter as electrospinning ($O(d) \sim 100$ nm) media using Nylon 6 and Polylactic acid (PLA) with varied parameters analyzing the efficiency of the filters. The parameters taken under consideration are the concentration of the solution, flow rate and RPM (rotations per minute) for spinning these filters, collecting them on a spunbond layer. These materials are then compared with those made from other techniques such as electrospinning and melt-blowing for its morphology, filtration efficiency, pressure drop, and reusability. Finally, a duckbill mask is constructed using these filters are its performance as a filter medium is evaluated using Particle Filtration Efficiency Tester TSI 8130 and pass the N95 threshold set by NIOSH standard.

Digital Assets, Digital Threads, and Digital Twins for Rapid Commercialization of Medicines: A Reference Architecture and Implementation to Enable Industry 4.0

Pablo A. Rolandi, Cenk Undey

Today's industrial digital infrastructure is obsolete and fragmented. ELNs are expensive vaults with poor data engineering. PLMs are non-existent or they lack the breadth and depth expected of a true lifecycle management application. While Enterprise Data Lakes have had some success, they can be fragile and are not fully democratized to reach citizen data scientists. Project Portfolio Management may barely exist beyond glorified MS Excel/Project files. Data Science's maximum expression is Random Forests. Machine Learning (e.g., image analysis and NLP) has not reached mainstream use, and In Silico Modeling results may not exist beyond an obscure and fragile shared folder. High Throughput Experimentation's data is untapped due to the drag imposed by data wrangling. Ontologies are missing, rendering end-to-end analyses a mammoth undertaking. Reports, dashboards, may be more commonplace, but GxP validation can be a major burden. With this current landscape, the Industry 4.0 revolution seems a very daunting task even for the most daring digital innovators in the industry. Finally, cultural norms and organizational constraints can make the transformation even more challenging.

In this work we present a future-view rooted in Amgen's Process Development vast experience in this digital transformation space, and we share with the industrial community, regulatory agencies, and software vendors the many lessons learned. We envision a future of FAIR data assets that are curated and stewarded via effective data governance practices and processes. We imagine a digital thread that fetches these data assets from their sources, and changes them transactionally and programmatically with full traceability and scalability. We recognize that digital twins born out of machine learning and first principles modeling techniques will create cycle times near their theoretical minimum thanks to ubiquitous computation and augmented intelligence. We hypothesize a near-future where these connected assets, threads, and twins will revolutionize CMC authoring, quality processes, and enable major manufacturing and supply efficiencies. Finally, we identify a set of skills, methods, and practices to deliver this vision. This blueprint can be easily extended to other chemical process industries.

Water and Vapor Transport through Angstrom-Scale Pores in Atomically Thin Graphene Membranes

Piran Kidambi

The transport of waterthrough nanoscale capillaries/pores plays a prominent role in biology, ionic/molecularseparations, water treatment, breathable and protective applications. However, the transport behavior of water (liquidwater and water vapor) through nanoscale apertures remains elusive. Here, we use centimeter-scale atomically thingraphene membrane with angstrom-scale pores as a model to probe the transport of liquid water and water vapor, andreport on orders of magnitude differences in transport rates of water vapor and liquid water. We demonstrate that large-area atomically thin graphene membranes show an order of magnitude higher water vapor transport rate thancommercially available ultra-breathable protective materials, despite having significantly smaller pores (<6.6Å) thanthose commercial materials (e.g.,ePTFE with mean pore size ~210 nm), indicating their potential application for next-generation breathable and protective materials.

Topological Scientific Representations for Materials and Protein Property Predictions

Aditi Krishnapriyan

Machine learning has emerged as an attractive alternative to experiments and simulations for predicting properties from structure, such as that of materials or proteins. Usually, such an approach relies on specific domain knowledge for feature design: each learning target requires careful selection of features that an expert recognizes as important for the specific task. The major drawback of this approach is that computation of only a few structural features has been implemented so far, and it is difficult to tell a priori which features are important for a particular application. Moreover, these models often have their limitations, including the loss of long-range, global information. To address these challenges, we incorporate topological information into the learning process. We capture a summary of the topological features in a given material or protein structure through trainable neural network layers. We show that incorporating topological information greatly improves predictive performance compared to current standard descriptors on a range of structure-based prediction tasks.

Enhanced Electrochemical CO_2 Reduction Using DNA-Based Nanomaterials

Gang Fan, Ariel Furst

Carbon dioxide can be converted to high value chemical products. Small molecules have been demonstrated to catalyze electrochemical CO_2 reduction (ECR) process. To apply them in a controlled manner, these catalysts are required to be proximal to an electrode surface for the reaction to proceed. We developed a general method to readily place molecular catalysts on electrodes using DNA hybridization-based immobilization. DNA-catalyst complexes were synthesized through an optimized conjugation method and immobilized on screen-printed carbon electrodes using DNA hybridization. We found that this immobilization method improved the catalytic efficiency of molecular ECR catalysts for both an iron and a cobalt porphyrin-based catalyst. Importantly, DNA-modified catalysts demonstrated higher stability than their unmodified counterparts. Using this method, we could specifically control the distance between the catalytic sites to assemble DNA-catalyst nanostructures and elucidate the structure and function relationship of CO_2 electroreduction process. This method affords tunability and specificity, providing a new strategy to improve electrochemical CO_2 reduction efficiency.

Predicting Accurate Activation Energies for the Conversion of C,O, and H Containing Molecules over Metal Surfaces Using Machine Learning.

Kari E. Cordes, Daniel Hutton, Florian Goeltl

Obtaining accurate activation energies for surface reactions from first principles calculations requires experienced researchers to set up and run these calculations and is associated with a computational cost of about 10000 CPUh. At the same time, an in depth understanding of the performance or the computational design of heterogeneous catalysts requires the description of a complex reaction network and the calculation of a large number of activation energies. Therefore, an acceleration of activation energy calculations would be highly desirable.

In this contribution we develop a machine learning based method for the prediction of activation energies for reactions of C, O, and H containing molecules on metal surfaces. As a first step we generate a database of around 6900 surface reactions based on literature data. For each datapoint we assign a set of descriptors, which include the reaction energy, the change in chemical identity, molecular size, bonding of the molecules to the surface, the used methodology, and the surface metal and structure. We use this representation to apply a combination of Random Forest Regression, Support Vector Regression, and Gradient Boosted Regression, and our best approach leads to a Mean Absolute Error of 0.15 eV for the prediction of activation energies. We furthermore analyze the impact of the different descriptors on the quality of prediction and find that the reaction energy and change of chemical identity are most important, whereas other descriptors, such as the surface structure, the used functional, or the used electronic structure code hardly influence the quality of predictions.

The machine learning based approach presented here is fast, easy to use, and accurate and is an enabling technology for an efficient screening of the performance of metal catalysts.

Insights into the Impact of Anion Exchange Membrane on the Electrochemical Reduction of CO_2

Bjorn Hasa, Ahryeon Lee, Feng Jiao

The electrochemical carbon dioxide reduction (eCO_2RR) into value-added chemicals and fuels offers a sustainable route to store the renewable sources' excess energy into chemical bonds and to close the anthropogenic carbon cycle. The CO_2 electroreduction has been studied in different reactor designs, such as H-type cell, flow rate cell, and membrane electrode assembly (MEA) configuration. The latest reactor minimizes both ohmic losses and mass transport limitations. An MEA configuration consists of two electrodes (anode and cathode) divided by a polymer membrane. Typically, eCO_2RR studies employ reactant reduction in the cathode and oxygen evolution reaction in the anode. The role of the membrane is to facilitate the ion transport between the two electrodes and chemically isolate the occurring electrochemical half-reactions. It should be noted that investigators have conceptualized the use of ion-exchange membranes in order to improve the reaction energy efficiency and rate. ***However, the impact of membrane properties on cell performance remains unexplored***.

Herein, we investigated the role of the anion exchange membrane on the CO_2 reduction performance using a membrane electrode assembly configuration. The experiments were carried out in a two-compartment cell with an aqueous anode feed ($KHCO_3$) and a gaseous cathode feed (50 ml min^{-1}). The cathode electrode was synthesized by depositing Cu nanoparticles (25 nm) into a gas diffusion layer (carbon paper, Sigaret 39).

This talk will show for the first time how the ion-exchange membrane affects the electrocatalytic activity of CO_2. Membrane properties have the ability to determine product crossover, product selectivity, and stability. We will discuss: i) the impact of membrane backbone structure and functionalization on the cell stability, ii) the analysis of anodic and cathodic outlet streams, iii) the transport of the charged species, and iv) the role of the membrane on product selectivity.

Deep Learning Assisted Investigation of Field-Dipole Interactions on Catalytic Ammonia Synthesis

Mingyu Wan, Han Yue, Jaime Notarangelo, Hongfu Liu, Fanglin Che

Currently, producing ammonia under mild conditions using various catalysts has obtained much progress but still has the following limitations: (1) electrochemistry yields only trace amounts of ammonia at modest current efficiencies due to the kinetically restricted initial N_2 dissociation step at room temperature; (2) biomimetic routes suffer from mass-transport limitations; (3) chemical looping is in the exploratory stage; and (4) non-thermal plasma suffers from fast energy loss and low energy efficiencies. A potential strategy to overcome kinetic limitations and enhance energy efficiency of ammonia synthesis at a modular setup is to apply a large electric field.

External electric fields (> 0.1 V/Å) can modify binding energies of reactive surface species and enhance catalytic performance of heterogeneously catalyzed reactions. In this study, we used DFT calculations—assisted and accelerated by a deep learning algorithm—to investigate the extent to which catalyzed ammonia synthesis would benefit from application of such external electric fields. This strategy allows us to determine which electronic properties control a molecule's degree of interaction with external electric fields. Our results show that: (1) a positive field promotes ammonia synthesis by lowering the overall energetics and decreasing the activation barriers of the potential rate-limiting steps (e.g., NH_2 hydrogenation) over Ru; (2) a positive field (> 0.6 V/Å) favors the reaction mechanism by avoiding kinetically unfavorable N≡N bond dissociation over Ru(1013). The deep learning algorithm developed here accelerates field-dependent energy predictions with acceptable accuracies by five orders of magnitudes compared to DFT alone and has the capacity of transferability, which can predict field-dependent energetics of other catalytic surfaces with high quality performance using little training data (***Figure 1***).[1]

Reference

1. M. Wan, H. Yue, J. Notarangelo, H. Liu, F. Che*, "Deep-Learning Assisted Electric Field-Accelerated Ammonia Synthesis", ***JACS Au***, Accepted, **2022**, ACS Editor's Choice.

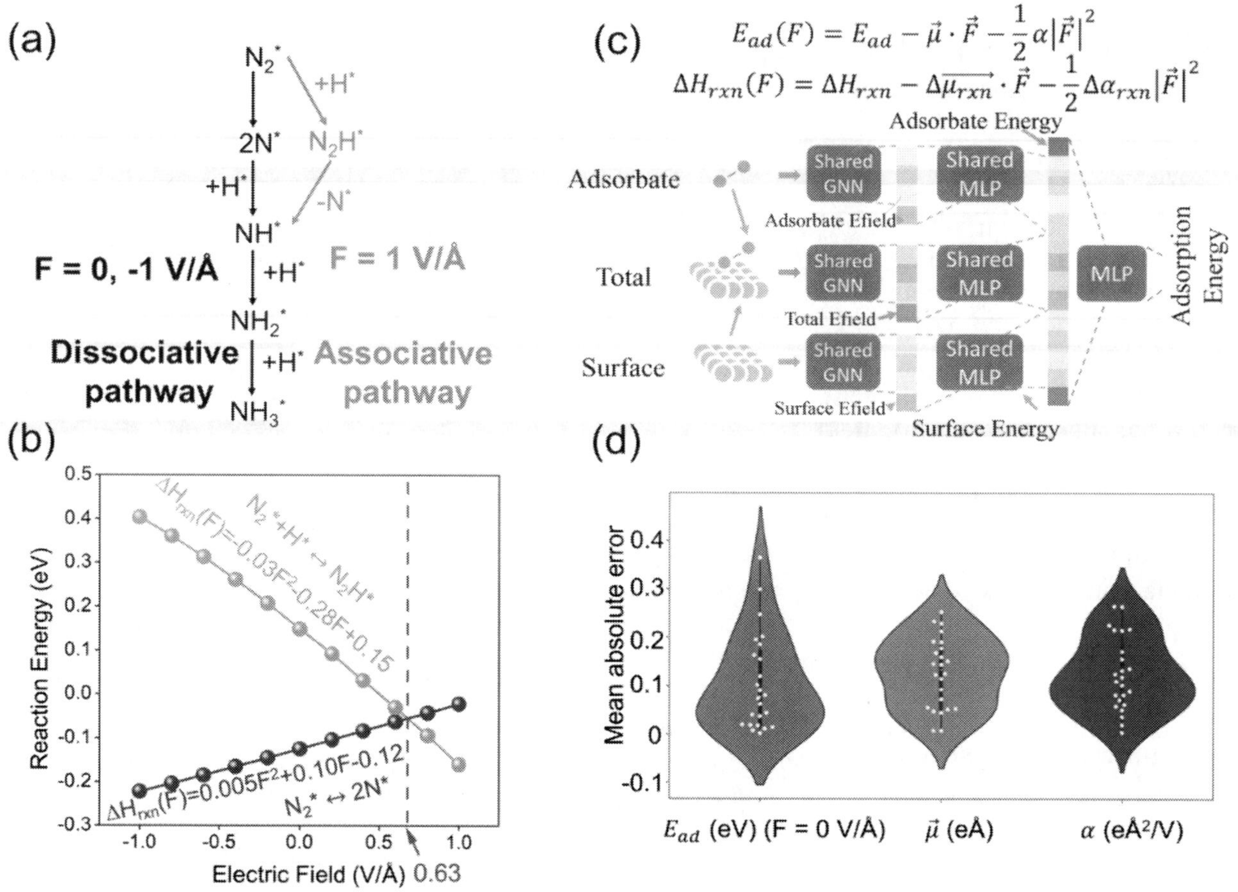

Figure 1. (a), (b) The positive field enhances ammonia synthesis via lowering the energetics of dinitrogen to ammonia and favors the reaction mechanism by avoiding kinetically unfavorable N≡N bond dissociation. (c) and (d) Deep-learning model to predict field-dependent adsorption.

A Novel Exergy Analysis Based on First and Second Law of Thermodynamics

Atsushi Tsutsumi

Energy consists of two different components: exergy and anergy, which are defined as the maximum work extractable from energy and the thermal energy at the standard temperature of 25°C, respectively. In the energy transfer process exergy is destructed with the entropy generation due to the energy transfer resistance and transformed into anergy, but anergy can be never re-transformed to energy. This induces the inequivalence between work and heat and the irreversibility of the process due to the entropy generation issued from the Second Law of Thermodynamics. So far, the energy conversion efficiency of energy process such as heat engine has been analyzed based on the First Law of Thermodynamics by means of the classical thermodynamic identity and the energy conversion equation in the scalar notation and pressure – volume (*P-V*) diagram which cannot express the exergy destruction as well as anergy. In this presentation, an exergy analysis based on the First and Second Law of Thermodynamics has been developed by using the thermodynamic identity and the energy conversion equation in the vector notation for energy with exergy and anergy and the temperature – entropy (*T-S*) diagram which can express the inequivalence between work and heat and the irreversibility of the process due to the entropy generation. This exergy analysis is adopted to elucidate the thermodynamic mechanism of open cycle process for irreversible heat engine with non-isentropic compression and expansion.

In terms of the modularization which excludes the exergy destruction caused by the irreversibility of adiabatic compression and expansion from a material flow process, material and energy production system of cycle process such as heat engine can be decomposed into the material flow cycle process with material stream which consists of a series of thermodynamic elementary processes (isentropic compression and expansion processes and isobaric heating and cooling processes) and the energy conversion modules with energy stream for the work and heat transactions of the material flow cycle process. In the energy conversion modules of energy transfer, the exergy destruction takes place with the entropy generation due to the resistance to mass and heat flow caused by pressure and temperature gradients, transforming exergy into anergy. On the other hand, the thermodynamic process in which the transition of equilibrium state is caused by changing the internal energy due to the work and heat transactions of the system with its surroundings has no exergy destruction of the internal energy because of no pressure and temperature gradients in the system at thermodynamic equilibrium state. Therefore, both exergy and anergy of the internal energy in the material flow cycle process are conserved. Thus, the anergy of output heat is the sum of that of input heat and the exergy destructions in the energy conversion modules for the non-isentropic compression and expansion. Then, we can drive the energy conversion equations for reversible and irreversible heat engine systems in vector notation. The output heat can be calculated from the sum of anergy of input heat and the exergy destructions in the energy conversion modules by use of the T-S diagram. This also gives the net output work extracted from the input heat and the actual thermal efficiency of heat engine system.

Selective Electrochemical CO_2 Reduction to Formate over Bismuth Nanosheets Derived By *in-Situ* Morphology Transformation of Bismuth Oxides

Jungkuk Lee, Hengzhou Liu, Yifu Chen, Wenzhen Li

The electrochemical CO_2 reduction reaction (CO_2RR) has attracted enormous attention as a promising technology not only to mitigate CO_2 accumulation, but also to convert CO_2 into valuable chemicals and fuels Formate/formic acid is a value-added product from CO_2RR that is an important intermediate for chemical industries such as medicines, rubbers, and leather. Bi-based catalysts have been studied intensively because of their low toxicity, environmental benignity, low cost, and potentially high durability. Bismuth nanosheets (BiNSs) have been recognized as a promising catalyst for electrochemical CO_2 reduction (CO_2RR) to formate, but its preparation typically involves an elaborate synthesis of Bi-precursors under elevated temperature and pressure. Here, we demonstrate a simple surfactant-free method of preparing Bi_2O_3-derived BiNSs (OD-BiNSs) by aqueous precipitation and cyclic voltammetry (CV) under ambient conditions. *In-situ* morphology transformation from Bi_2O_3 to BiNSs was observed during CV, in which the presence of oxygen and the initial morphology of Bi_2O_3 are crucial to the phase transformation. The as-prepared OD-BiNSs showed 93% of faradic efficiency (FE) to formate with a partial current density of 62 mA cm^{-2} at -0.95 V_{RHE} in the H-cell, and >94% of FE at 50–200 mA cm^{-2} with cell voltages of 2.4–4.0 V in the flow cell. Our recent progress in CO2 reduction will also be presented.

Exploring the Detrimental Effect of Water Activity on Liquid Phase Fatty Nitrile Production for in-Situ Water Removal

Carola Raffel, Vishnu Prasad Sureshkumar, Melissa Assirelli, Jacob van den Berg, Antoon ten Kate, John van der Schaaf

To help mitigate the effects of climate change, the chemical industry needs to become more sustainable by using raw materials from renewable resources and by intensifying existing processes to become more energy and material efficient. This is also the case for the production of fatty nitriles, which are important chemical intermediates, and that can be produced from renewable resources [1]. To produce fatty nitriles, fatty acid and ammonia are reacted to form fatty amide while producing a stoichiometric amount of water. The fatty amide in turn dehydrates to give the fatty nitrile and an additional stoichiometric amount of water as seen in figure 1A. That means that for every mole of nitrile, two moles of water are produced. As both reactions are equilibrium reactions, see figure 1A, removal of water is necessary to increase the nitrile yield and bring the reaction to completion. In industry, this is achieved in a semi-batch reactor with a high ammonia excess that sweeps out the water from the liquid phase, which may not be the most sustainable process operation as it requires a significant recycle of ammonia and associated process and energy costs.

To intensify the nitrilation reaction, a system that combines a reactor with a separation unit is beneficial. In this reactor, the water produced during the reaction can be removed in-situ to push the conversion beyond what is expected in a conventional reactor. Examples for this type of intensified reactor are membrane reactors, reactive distillation, or trickle bed reactors. To be able to choose the reactor configuration most beneficial for intensifying the nitrilation reaction, a complete model of the reaction needs to be available: the intrinsic kinetics of the reaction, the mass transfer phenomena as well as the thermodynamic behavior of the system to describe its gas-liquid interactions. Key for the latter is the modified Raoult's law, assuming the two phases are in equilibrium, as seen in equation 1:

$$p * y_i = \gamma_i * x_i * p_i^{sat} \quad (1)$$

Where γ_i is the activity coefficient of component *i*, y and x the gas and liquid phase mole fraction and p_i^{sat} the saturation pressure of component *i*.

The activity coefficients of water and ammonia in the fatty compounds are not readily available from literature since their measurement at reaction conditions is complicated by their reaction with the fatty compounds. Therefore, the activity coefficients are simulated to allow a characterization of its gas liquid behavior and through that an informed choice for an intensified reactor.

The reaction system considered in this work consists of water, ammonia, oleic acid (C18:1), oleic amide and oleic nitrile. At the typical reaction temperatures of 250-300°C [1] and reaction pressures of 1-8 bar, the vapor pressures of the oleic compounds are negligible. Therefore, their vapor fractions are considered as zero and only the vapor-liquid behavior of water and ammonia needs to be known.

In this work, the activity coefficients of ammonia and water in oleic acid, oleic amide and oleic nitrile are determined by simulation. The coefficient of ammonia in oleic nitrile could be measured experimentally since the two compounds do not react with one another. At a range of temperatures in a batch reactor known amounts of nitrile and ammonia were introduced, the reactor was heated, letting an equilibrium establish itself, noting the pressure. Using the modified Raoult's law, the activity coefficients could be determined. As the other combinations of water and ammonia with oleic acid, oleic amide and oleic nitrile react with each other, their activity coefficients were simulated in COSMO-RS, a quantum chemistry-based thermodynamics model. All molecules were calculated with the GGA:PBE functional and the QZ4P basis set in AMS-ADF.

Figure 1B shows the comparison between the experimentally determined activity coefficient of ammonia in oleic nitrile and the simulated activity coefficient of ammonia in nitrile using COSMO-RS. As can be seen, model and experiment are in good agreement in the range of 100-150°C. It also shows that the simulated behavior of ammonia in nitrile at lower temperatures is not ideal ($\gamma NH3>1$) and approaches ideal behavior at reaction temperatures. This was also found for the activity coefficients of ammonia in oleic acid and oleic amide.

The simulated activity coefficients of water in oleic acid, oleic amide and oleic nitrile are shown as a function of temperature in figure 1C. At lower temperatures, the activity coefficients are higher while they decrease drastically when approaching reaction temperatures. Furthermore, the differences of the activity coefficient of water in oleic acid, oleic amide and oleic nitrile are highlighted: oleic nitrile as an aprotic molecule repulses water the most, while the protic oleic acid and oleic amide interact more favorably with water.

These simulation results show that ammonia at reaction temperatures can be treated as behaving ideally. On the other hand, water has an activity coefficient of around 4 at reaction conditions which is drastically lower than at ambient conditions. This shows that the inline water removal of the intensified reactor needs to be efficient to counteract this behavior. With the data presented in this work, a comparison between different intensified reactor types such as membrane reactors, reactive distillation, or trickle bed reactors will be carried out. Their feasibility and effectivity will be judged with regards to space-time-yield of nitriles, amount of water removed as well as amount of ammonia excess necessary for full conversion. This allows the choice for the intensified reactor type most promising for a more sustainable production of fatty nitriles.

References

[1] A. Mekki-Berrada et.al., 'Ammoniation-dehydration of fatty acids into nitriles: Heterogeneous or homogeneous catalysis?,' ChemSusChem, vol. 6, no. 8, 2013.

Figure 1A

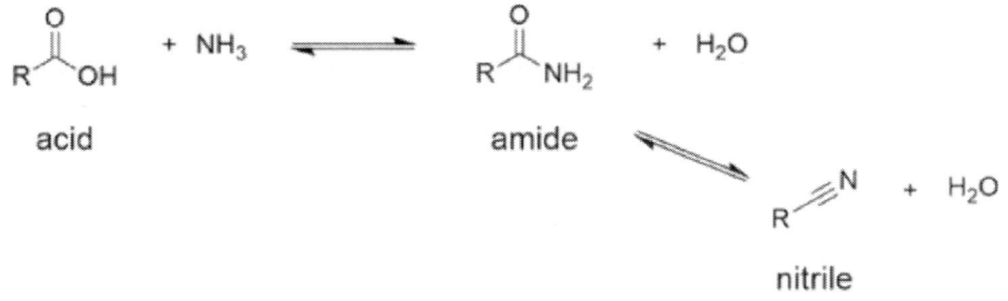

Figure 1B

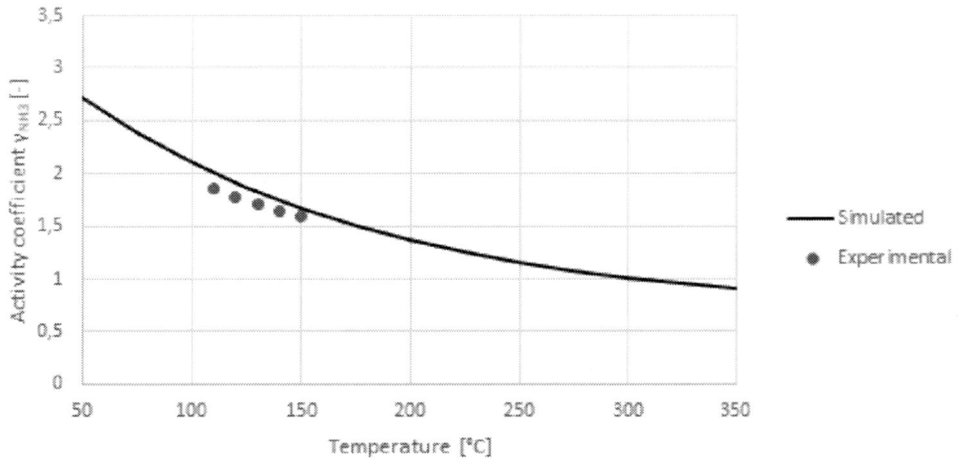

Figure 1C

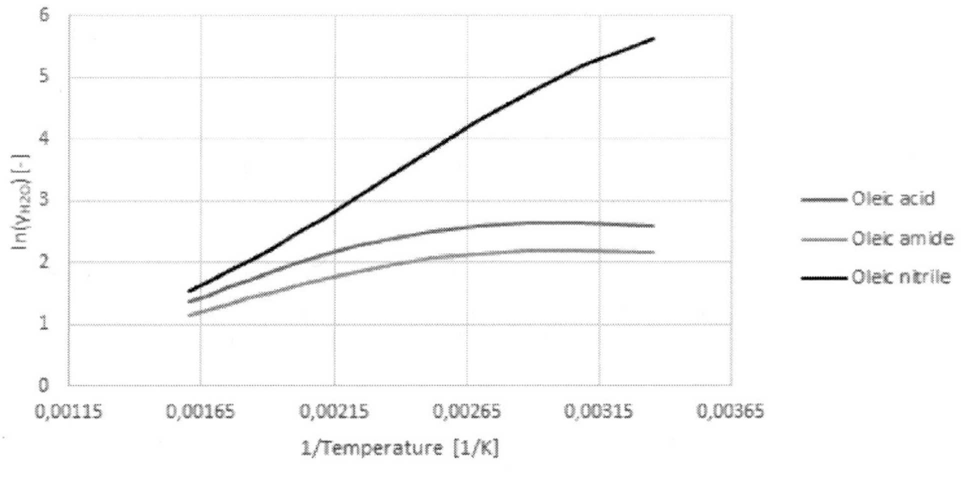

Process Modularity Guides Process Intensification - Application to Safety

Arick Castillo, Aburto Jorge, Jhuma Sadhukhan, Elias Martinez Hernandez

Process intensification is a tool that has drawn attention in recent years. Several efforts have shown that implementing hybrid units such as divided-wall distillation columns or reactive distillation columns, because of the union of two or more units, allows improving profitability, environmental impact and safety of a chemical process. A methodology to intensify a chemical process guided by the principle of modularity is presented in this paper. A process network is decomposed into modules by applying a community detection algorithm to find the process units to be integrated into an intensified 'module' in order to improve the Fire and Explosion Damage *Index (*FEDI). A case study involving the separation process of the ethanol-butanol-water mixture is used to show the usefulness and practicality of using the principle of modularity for intensification, leading to more inherently safe processes. The results showed that the safety indicator improves when the process is intensified guided by the modularity value, in comparison to the non-intensified process and an empirically intensified alternative.

Rheological Protocol to Assess Latex Coagulation Dynamics for the Thin Glove Coagulant Dipping Process

Monday Okoronkwo

Many products that directly impact the quality of human life today, including gloves, catheters, condoms, and baby bottle teats, are made through latex-dipping technology. While a variety of methods have been developed – e.g., particle counting, turbidimetry, microscopy, and light scattering – which are suitable for studying the coagulation of latex at very low concentrations, little or no progress has been made in developing methods suitable for in-situ characterization of latex coagulation in concentrated solutions (e.g., the system used in the dipping process). This paper presents an original rheological protocol for assessing and optimizing latex coagulation dynamics for the thin glove coagulant dipping process. The method involves, first, the use of small amplitude oscillatory rheology to characterize the time-dependent evolution of the viscoelastic properties (e.g., dynamic moduli and phase angle), to quantify the coagulation kinetics. Second, the percolated latex film (after coagulation) is assessed by compressional rheology to evaluate the thickness and elastic modulus. The protocol is herein demonstrated using a commercial-grade carboxylated butadiene-acrylonitrile copolymer (XNBR) latex and nitrate-based coagulant solutions containing various counterions such as calcium, magnesium, aluminum, sodium, and ammonium ions. The results show that the method is sensitive to detect small changes in ionic speciation, the concentration of coagulant formulations, and small changes in temperature. Hence, it can be suitably applied to assess and optimize latex-coagulant formulations, and processing parameters for the latex dipping process.

Early-Age Properties of Fly Ash Blended Calcium Sulfoaluminate Cement Pastes

Sukanta K. Mondal, Monday Okoronkwo

Calcium sulfoaluminate (CSA) cements are a promising low CO_2 alternative to commonly used ordinary Portland cements (OPC). This study aimed to utilize a low-priced, most abundant industrial solid waste, fly ash, to make the CSA cement systems more sustainable. The effect of fly ash on the early-age properties of a CSA-based cement paste was investigated. The influence of Class C (FA-C) and Class F fly ashes (FA-F) on the rheological behavior and hydration kinetics of CSA cement were studied. Three rheology-based workability parameters, i.e., (i) the rate of loss of flowability **($\mathbf{m_1}$)**, (ii) the rate of hardening **($\mathbf{m_2}$)**, and (iii) placement limit **($\mathbf{t_p}$)** were characterized for the fly ash-dosed CSA pastes. The effect of both fixed water-to-cement (w/c) and water-to-binder (w/b) ratios were examined. At fixed w/c, with increasing fly ash contents of up to 15 wt%, the **$\mathbf{m_1}$ and $\mathbf{m_2}$** increased, while the **$\mathbf{t_p}$** decreased. In contrast, at fixed w/b, a slightly opposite trend was observed, indicating the fly ash addition improved the workability of the pastes. Compared to FA-F modified systems, FA-C modified systems are generally found more reactive due to its high Ca content. The results indicate that increasing fly ash content accelerates the hydration of the calcium sulfoaluminate cement pastes, chiefly due to filler effects than pozzolanic effects, which is attributable to non-availability of free portlandite in the pore solution, especially in the case of FA-F modified pastes. The addition of fly ash slightly reduced the 24-h compressive strength, and because of the less reactivity, the reduction of strength is more in FA-F systems. Both the characteristic rheological behavior and the heat of hydration agree to a domination of filler effect in the hardening process. Thermodynamic simulations of the phase assemblage indicate an increase in the volume of ettringite and strätlingite, leading to reduced porosity with increasing degree of fly ash reaction. Overall, the formation of ettringite is stabilized more in FA-F systems. This study provides some fundamental insights on the interaction between fly ash and CSA cement and encourages further studies towards the establishment of standards on the increase of effective usage of fly ash in cement systems, especially in CSA cement systems.

Graduate Student Award Session: Organic-Inorganic Interactions in 3D Printed Bouligand Nanostructures

Mohsen Esmaeili, Kyle George, Nasser Nikfarjam, Nader Taheri-Qazvini, Monirosadat Sadati

Due to the unique functionality of hierarchical structures in nature, researchers have attempted to mimic complicated biological structures. Among these complex hierarchies is the concentric cylindrical chiral nematic structure of collagen fibers in cortical bone, known as Bouligand structure, from which the remarkable compression strength arises. Cellulose nanocrystals (CNCs) can form the chiral nematic structure in aqueous suspensions, thereby can be a candidate as an organic component to resemble the concentric chiral structure. The high compression strength, however, arises from the presence of inorganic minerals, as widely observed in many biological systems. Here, we perform the mineralization of organized CNC's chiral nematic structures in the axially aligned concentric configuration by using direct ink writing (DIW), as a 3D printing technique. The formation of such a complex structure can be expected when CNC inks with the shear-induced para-nematic structure, formed through the DIW's nozzle, leaves the DIW's nozzle and the chiral nematic structure start to recover within the cylindrical confinement. To immobilize the recovered chiral nematic structure during DIW, a mixture of photo-curable monomers of Acrylamide (AAm) and Acrylic acid (AAc) are incorporated into the aqueous CNC suspensions at the CNC to additive (AAm+AAc) ratio of 1:4. Polarized optical microscopy (POM) confirms that incorporating photo-curable additives at this ratio does not disturb the chiral nematic structure of CNC particles. However, the final reactive chiral nematic ink exhibits liquid-like behavior and does not provide self-sustainable 3D printed filament. Here, we employ the embedded 3D printing approach, where jammed environments, like Carbopol microgels, provide spatial support for extruded inks while UV light advances photo-polymerization. The uniform alignment of concentric chiral nematic structures in free-standing hydrogels after photo-polymerization is confirmed by POM and electron microscopy. To induce the mechanical strength, the mineralization is performed by immersing and swelling free-standing hydrogels in calcic solutions, where Calcium ions precipitate within hydrogels. The degree of mineralization, however, depends on the swelling ratio and anionic groups of CNC and AAc (sulfate, and carboxyl groups, respectively). At a constant ratio of CNC:additives, increasing the ratio of AAc over AAm results in both a higher swelling ratio and more anionic groups in the systems, leading to a higher degree of mineralization, confirmed by elemental identification by energy-dispersive X-ray spectroscopy (EDS), and thermogravimetric analysis (TGA). This experiment suggests an approach for fabricating hybrid material with axially aligned concentric chiral nematic structure to obtain outstanding mechanical properties inspired by complex biological systems.

Graduate Student Award Session: Shear-Induced Optical Properties in Photonic Hydroxypropyl Cellulose Solids

Kyle George, Mohsen Esmaeili, Nader Taheri-Qazvini, Monirosadat Sadati

At concentrations ranging from 50 to 70 wt% in water, Hydroxypropyl cellulose (HPC) self-assembles into chiral nematic liquid crystalline (cholesteric) phase with characteristic helical (pitch) spacing on the scale of visible wavelength exhibiting intriguing chiral photonic property. The photonic and optically responsive properties offered by such solutions are attractive as potential substitutes for pigments and dyes as well as use in biosensing applications. However, preservation of both the cholesteric pitch length and orientation have proven to be challenging. During the drying process, HPC's lyotropic behavior causes the pitch to compress to length scales on the order of ultraviolet light, resulting in the loss of the photonic properties developed in the solution. The production of structurally colored solids from these chiral solutions can be achieved by introducing a compatible additive molecule, such as low molecular weight polyethylene glycol (PEG), which is capable of forming hydrogen bonding networks within the cholesteric structure. The development of this physically cross-linked system provides support to the cholesteric pitch as solvent is removed from the system, resulting in structurally colored solids which span the visible light spectrum. Furthermore, 3D printing techniques such as direct ink-writing (DIW) can be used to control the orientation of the cholesteric self-assembly in the final solid. Rheological characterization of these solutions allows for the determination of an optimal shear rate (23 s^{-1}) to be applied during the DIW process, with solids obtained from these shear conditions displaying uniform coloration across the entire material. The optical properties of the film can be further manipulated by exposing solutions to shear rates which well exceed that of the optimal shear rate (approximately 100 s^{-1}). Solids obtained from such processing conditions display colors consistent with films obtained using optimal shear rates, however, exposure to high shear rates results in a tilted cholesteric structure with strong, viewing angle dependent, color. Modification of the optical properties of HPC based solids through the imposition of shear forces provides a simple method for the production of interactive materials.

Enhancement of Thermal and Hydraulic Characteristics of Microchannel Heat Sinks By Employing sCO_2 As a Coolant

Abdallah S. Berrouk, Muhammad Saeed

The supercritical carbon dioxide (sCO2) operating close to its critical point can be of great value because of its promising thermophysical properties. Hence, the thermohydraulic performance of the sCO_2-cooled microchannel heat sinks (MCHS) is investigated numerically and compared with the conventional water-cooled MCHS. To perform the conjugate heat transfer calculations mentioned above, a 3D Reynolds Averaged Navier–Stokes (RANS) model is established and validated for both coolants, and the shear stress transport (SST) turbulence model is used to model the flow turbulence. Thermophysical properties of sCO_2 are incorporated via a high-resolution real gas property (RGP) file to capture sharp changes in its properties. The comparison of the performance of the MCHS with two coolants reveals that sCO_2-cooled MCHS is up to 32% higher in comparison with water-cooled MCHS. Further, base temperature with the former was found 8°C less than the latter. It is also found that the sCO_2-cooled MCHS can withstand the off-load heating loads without any change in the base temperature. Moreover, it is found that the sCO_2-cooled MCHS reduces the pressure drop up to 7 times compared to water-cooled MCHS, while the friction factor for the proposed coolant is up to 10 times smaller than the water-cooled MCHS. Finally, based on overall performance evaluation criteria performance of sCO_2-cooled MCHS was found up to 2.2 times better than the water-cooled MCHS.

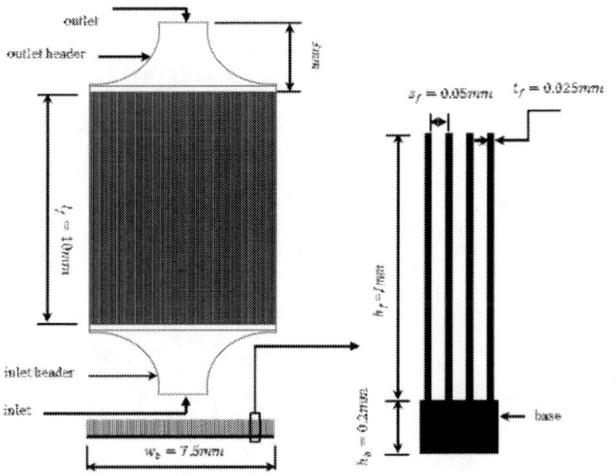

DBD Plasma Reforming of Diverse Hydrocarbon Feedstock: Catalytic Experiments, Process Design and Life Cycle Analysis

Guoqiang Cao, Yue Xiao, Chien-Hua Chen, Jonas Baltrusaitis

Catalytic dry methane reforming as well as ethanol reforming was investigated using DBD plasma reactor. The presentation will summarize results obtained during catalytic runs with conversion and selectivity obtained, the decoking experiments, temperature and catalyst effect. Further, tri-reforming experiments facilitated by DBD plasma were performed where water vapor was used to decrease the excess of CO_2 needed for efficient single pass CH_4 conversion. The cylindrical reactor design will be presented that increased the overall conversion at specific energy input without additional heating or catalyst. These laboratory data were used to construct a process model in silico to process CO_2 produced via conventional distributed small scale H_2 production process and the resulting economics as well as environmental impacts were calculated and presented. Finally, bioethanol DBD plasma reforming was performed to calculate product distribution with and without catalyst, together with the proposed technological process as well as the corresponding environmental impacts via LCA analysis.

Supplying By-Products to Distributed Chemical Processes for Sustainability

Thérèse G. Lee Chan, David A. Janes

Enterprise Input-Output (EIO) models were used to analyze the flows of resources, such as energy and materials, around petrochemical-manufacturing networks. From this analysis, it was found that the Point Lisas Industrial Estate exhibits the rudimentary characteristics of an eco-park and that there are various opportunities for industrial symbiosis (IS). To accelerate the intensification of the IS, multi-criteria decision support tools were developed utilizing multi-objective optimization (MOO) algorithms on a MATLAB platform. EIO model-based case studies were undertaken to consider the sharing of waste streams (e.g., contaminated water) and valuable by-product streams (e.g., process carbon dioxide) through reuse between different petrochemical plants in the network. Various MOO techniques were used to minimize the release of pollutants from plants and the associated implementation costs. As the modelled networks became more complex, analysis of the obtained solutions showed that bespoke code using scalarized methods, like the ε-constraint method, were more robust than other methods relying on the packaged MOO functions available in MATLAB for models with linear constraints. The results indicated that, if promoting the IS attitude led to more collaboration amongst companies and exploitation of new manufacturing opportunities, such as a mineral carbonate plant, then triple bottom-line benefits (e.g.: reduced carbon dioxide emissions, new revenue streams and skilled jobs) and increased materials circulation might be realized.

Tailoring Nanoporous Graphene Via Machine Learning: Predicting Probabilities and Formation Times of Arbitrary Nanopore Shapes

Rahul Sheshanarayana, Ananth Govind Rajan

Nanopores in graphene, a 2D material, are currently being explored for various applications, such as gas separation, water desalination, and DNA sequencing. The shapes and sizes of nanopores play a major role in determining the performance of devices made out of graphene. However, given an arbitrary nanopore shape, anticipating its creation probability and formation time are challenging inverse problems, solving which could help develop theoretical models for nanoporous graphene and guide experiments in tailoring pore sizes/shapes. In this talk, I will discuss the development of a machine learning (ML) framework to predict these target variables, based on data generated using kinetic Monte Carlo simulations and chemical graph theory. Thereby, we enable the rapid quantification of the ease of formation of a given nanopore shape in graphene via silicon-catalyzed electron-beam etching and provide an experimental handle to realize it in practice. We use structural features such as the number of carbon atoms removed, the number of edge atoms, the diameter of the nanopore, and its shape factor, which can be readily extracted from the nanopore shape. We show that the trained models can accurately predict nanopore probabilities and formation times with R^2 values on the test set of 0.97 and 0.95, respectively. Not only that, we obtain physical insight into the working of the model and discuss the role played by the various structural features in modulating nanopore formation. Overall, the work presented in this talk will provide a solid foundation for experimental studies to manipulate nanopore sizes/shapes and for theoretical studies to consider realistic structures of nanopores in graphene.

Prediction of Propagation Rate Constants of Acrylate-Based Polymers Using Multi-Input-Multi-Output Graph Attention Network

Tung Nguyen, Mona Bavarian

Acrylate-based polymers are commonly employed in electronics applications, such as anti-reflective coatings and photoresists. These polymers constitute essential components in the formulation of these mixtures. Thus, precise manufacturing and quality control of both the chemical and physical properties of the polymer resins are essential to satisfy the need for development of high-precision coatings. A mathematical model is an indispensable tool to predict and control the properties of polymers, and it can be implemented in manufacturing systems for real-time control of reactions. Along these lines, kinetic modeling is central to developing optimal reaction conditions. Among different parameters, propagation rate constants play an essential role in determining molecular weight (MW), molecular weight distribution (MWD), and composition for copolymers, thus influencing the performance of products. Therefore, accurately determining the rate constants of monomers is critical in modeling polymerization processes, product design, and manufacturing scale-up. As Machine Learning (ML) has become popular due to its ability in adapting complex mathematical calculations to big data and decision-making, this tool is aggressively utilized in the polymer field. Also, a combination of multiple inputs in models has become popular in the ML field because they significantly improve the predictive capability of models. Considering the potential of the graph-based models, GAN is utilized in this research. Along with GAN, we developed the Multi-Input-Multi-Output Graph Attention Network (MIMO GAN) model to improve the prediction of kinetic rate constants. The predicted rate constants were then compared with the experimental data. We also compared the prediction capability of the Multi-Input-Multi-Output GAN (MIMO GAN) model against the Single-Input-Multi-Output GAN (SIMO GAN) model. We found that the MIMO GAN model outperformed the SIMO GAN model and resulted into the improvement of R^2 value from ~ 50 to 90% (i.e., R^2 value of 89 and 91% for k_{11} and k_{22}, respectively).

Model-Based Design and Dynamic Analysis of Piston-Type Work Exchanger for Mechanical Energy Recovery

Abdurrafay Siddiqui, Yinlun Huang

Work exchanger network (WEN) synthesis has drawn attention in recent years, as optimal design of a WEN can recover mechanical energy with significant economic benefits in process plants. A recent trend is to integrate both work exchange and heat exchange in a network system. In these efforts, the process unit used in the WEN is key. One type of units is the piston-type direct work exchanger. A successful industrial application of this type of units is in seawater desalination plants, but the process streams are in liquid phase. In recent years, we have been investigating a type of direct work exchanger where streams in gas phase perform work exchange as there exist various pressurized process streams in gas phase in chemical plants. There are a number of technical challenges in designing and operating such types of direct work exchangers.

In this paper, we introduce a fundamental modeling method for characterizing work exchange dynamics under different operational scenarios, such as in adiabatic, isothermal, and non-isothermal conditions. The model-based simulation enables the understanding of dynamic behavior of work exchange and thus provides guidelines for unit design and control. We implement the models using Aspen Plus, where ideal and non-ideal gas streams in different operating conditions and heat integration feasibility are studied. Process dynamics analysis provides valuable information for operational control and unit design. In the end, we will also present a case study on WEN design and operation with economic and safety analysis.

Ionic MOF-Cellulose Acetate Electrospun Membranes for Selective Removal of Selenium from Water.

Sweta Modak, Atta Ur Razzaq, Milad Esfahani

Selenium shows the tendency of bioaccumulation in the food chain and shortens the life span of aquatic and wild animals. The concentration of selenium >0.05ppm is hazardous for human health, and hence is crucial to remove it from drinking water. Metal organic frameworks (MOFs) are gaining significant attention recently as adsorbents due to their high porosity, adsorptivity and selectivity. In this study, the nickel-based ionic MOF (Ni-iMOF) was synthesized via solvothermal using $NiSO_4.8H_2O$ and tris(4-(1H-imidazol-1yl)phenyl)amine as ligand. Ni-iMOF can selectively adsorb selenate and selenite based on ion-exchange adsorption mechanism in the presence of competing ions, including SO_4^{2-}, CO_3^{2-} and HCO^{3-}. In addition, cellulose acetate (CA) shows adsorptive behavior towards selenium; however it lacks selectivity towards and in the presence of competing ions. We report here the synthesis, characteristics, and selective removal of selenite and selenite by hybrid electrospun CA-Ni-iMOF nanofibers. The effect of Ni-iMOF on the physicochemical properties of CA fibers, such as fiber diameter, and morphology, is investigated. In addition, different electrospinning conditions, including CA concentration (10-20 wt%), solvent mixture ratio N,N-dimethylformamide (DMF): Acetone (20:80, 50:50 and 80:20 (v:v), tip to collector distance (10-15 cm) , voltage (10-25 kV), solution feed rate (2-10 mL/h) and collector rotating speed (600-630 rpm) are used and the selective performance of the fabricated CA-Ni-iMOF are studied at different operation conditions (temperature (25-45°C), pH (2-12) and varying ionic strength (SO_4^{2-}, CO_3^{2-} and HCO^{3-})).

172a AI-Driven Drug Discovery and Manufacturing Using Automated Ontology-Based Information Extraction

Vipul Mann, Shekhar Viswanath, Shankar Vaidyaraman, Jeyakumar Balakrishnan, Jon Dieringer, Venkat Venkatasubramanian

The recent COVID-19 pandemic has highlighted the need to discover and manufacture drugs efficiently, quickly, and cost-effectively. This is a major challenge given the information complexity and overload that needs to be managed effectively in the discovery-to-delivery cycle. Computational tools and AI-based methodologies are crucial going forward to address this challenge. Here, we present a framework that could be used to automatically extract information from structured and unstructured documents with minimal supervision. These documents span several sources (briefing documents, research activities, electronic lab notebooks, process simulation documents) and different stages (research activities, experiments, manufacturing, sales, customer reports) of drug development. The AI-driven information extraction framework comprises the following components – ontologies, entity recognition, concept detection, and relation extraction. Ontologies offer an efficient means for organizing information hierarchically in the form of class-subclass relationships that are connected through object and data properties. Ontology-based information extraction is important because they preserve the semantics, incorporate domain knowledge in AI systems, and are flexible.

While there are several ontologies in the chemistry, biochemical, and medical domains, we present a novel and comprehensive ontology that captures different aspects of drug development and manufacturing including materials, materials properties, unit operations, the final drug product, the risks associated at different stages, and so on. This ontology is an extension of the previously reported Purdue Ontology for Pharmaceutical Engineering [1,2], the first comprehensive ontology in this domain. We integrated this ontology with a weak supervision approach for entity classification [3] that uses a BioBERT model [4]. We demonstrate the performance of the developed framework on several information extraction tasks using actual documents for commercial drugs and highlight the potential of such frameworks for accelerated drug discovery and manufacturing in the future.

<u>References:</u>

1. Leaelaf Hailemariam and Venkat Venkatasubramanian. Purdue ontology for pharmaceutical engineering: Part 1. Conceptual framework. Journal of Pharmaceutical Innovation, 5(3):88–99, 2010

2. Leaelaf Hailemariam and Venkat Venkatasubramanian. Purdue ontology for pharmaceutical engineering: Part 2. Applications. Journal of Pharmaceutical Innovation, 5(4):139–146, 2010.

3. Jason A Fries, Ethan Steinberg, Saelig Khattar, Scott L Fleming, Jose Posada, Alison Callahan, and Nigam H Shah. Ontology-driven weak supervision for clinical entity classification in electronic health records. Nature communications, 12(1):1–11, 2021.

4. Jinhyuk Lee, Wonjin Yoon, Sungdong Kim, Dong Hyeon Kim, Sunkyu Kim, Chan Ho So, and JaewooKang. Biobert: a pre-trained biomedical language representation model for biomedical text mining.Bioinformatics, 36(4):1234–1240, 2020.

172b Deciphering Longitudinal Optical-Density Measurements to Guide Antibiotic Use: A Model Based Approach

Iordanis Kesisoglou, Vincent H. Tam, Michael Nikolaou

Background: individualized design of drug administration for treatment of bacterial infections can benefit from experimental data on the effect of candidate drugs on infecting bacteria. In previous work, our group developed a model-based technique that renders usable optical density measurements of bacteria in suspension exposed to antibiotics over time, as these measurements count both dead and live bacterial cells and thus are practically useless. The fitted model can be used to design therapeutic treatment under clinically relevant pharmacokinetic conditions.

Objectives: Here we use in vitro simulation on a hollow-fiber infection model to validate predictions by the fitted model on the bactericidal efficacy of drugs that follow clinically relevant pharmacokinetics.

Methods: Optical-density measurements were taken over 20 hours in an automated instrument, based on which bacterial (*Acinetobacter baumannii*) cell counts over time were assessed under exposure to various time-invariant antibiotic concentrations. The model separated live from dead cell, thus estimating antibiotic kill rates at various concentrations. Subsequently, predictions were made on the bactericidal efficacy of the same antibiotic following a periodic profile of injection and exponential decline. These predictions were tested in an in vitro hollow-fiber infection model.

Results: Analysis suggested that suppression of the bacterial population could not be expected with clinically relevant ceftazidime concentrations alone. A combination of ceftazidime and amikacin (at a 2:1 ratio) would be necessary. These predictions were validated in the hollow-fiber infection model.

Conclusions: In vitro simulations substantiated the feasibility for clinicians to combine easily automated optical-density measurements with the proposed mathematical framework for the design of effective therapeutic treatments of challenging bacterial infections in a clinical setting. Additional validation experiments will be undertaken to consolidate the utility of the proposed method.

172c A Mathematical Model to Predict the Drug Release Profile in Bilayered Osmotic Controlled Release Tablets

Bhawana Tomar, Mahesh S. Tirumkudulu, Sweta Manthena, Weili Yu, Alfred Berchielli, Pankaj Doshi

The Oral Osmotic pump tablet, where the drug release is based on osmotic pressure gradient, has many advantages over other controlled drug delivery systems, namely, lower incidence of adverse reactions, better compliance in patients, decreased dosing frequency, independent of gastric pH, agitation, and hydrodynamic condition. One of the earliest osmotic pumps were the elementary osmotic pump or the extrudable core system (ECS), where a homogeneous mixture of drug, osmogen and polymer were coated with a semipermeable membrane. The advantages of ECS tablets are their ease of preparation and delivery of water-soluble drugs at near zero-order rates. At the same time, the ECS tablets are inefficient in delivering water-insoluble drugs due to the low osmotic pressure generated by the insoluble drug. The problem of incomplete extrusion is solved by using a bilayer system (push-pull system), also known as Swellable core technology (SCT), where a second layer of homogenous mixture of polymer and osmogen (sweller layer) is coupled to the drug layer. The sweller layer expands upon hydration, thereby providing additional pressure for the extrusion of the dispersed drug layer ensuring high percentage of drug release.

In this work, we model a **bilayer** osmotic controlled-release tablet to predict the drug release rate as a function of several key parameters. These parameters include the coating thickness and the tablet size as well as the ratios of excipients and active pharmaceutical ingredient. The model describes dynamically all the main release processes occurring during the dissolution of the tablet, namely, the solvent influx driven by the difference in osmotic pressure across the coating in both drug and sweller layer, dispersion of core components (drug and polymer), swelling of the tablet due to solvent accumulation, build-up of hydrostatic pressure inside the tablet, tensile stress acting on the coating, and the extrusion of the dispersed core components, dissolution of drug particles in bulk media, swelling of the polymer in the sweller layer, and pressure exerted by sweller layer onto the drug layer. The model is validated by comparing the predictions with drug release data for two drugs of varying solubility and for the same drug in different dissolution media. The drug release predictions match the measurements well, confirming that our model is well suited for elucidating the osmotically controlled drug release process in SCT tablets.

The attached figure presents the model predictions of drug release profile for the same drug but in two different dissolution media with different solubility. The figure plots (a) model predictions of drug release in dissolution media with high solubility, and (b) model predictions of drug release in water having low drug solubility. Presence of the sweller layer ensures that the drugs are extruded out at the same rate despite the large variation in solubility.

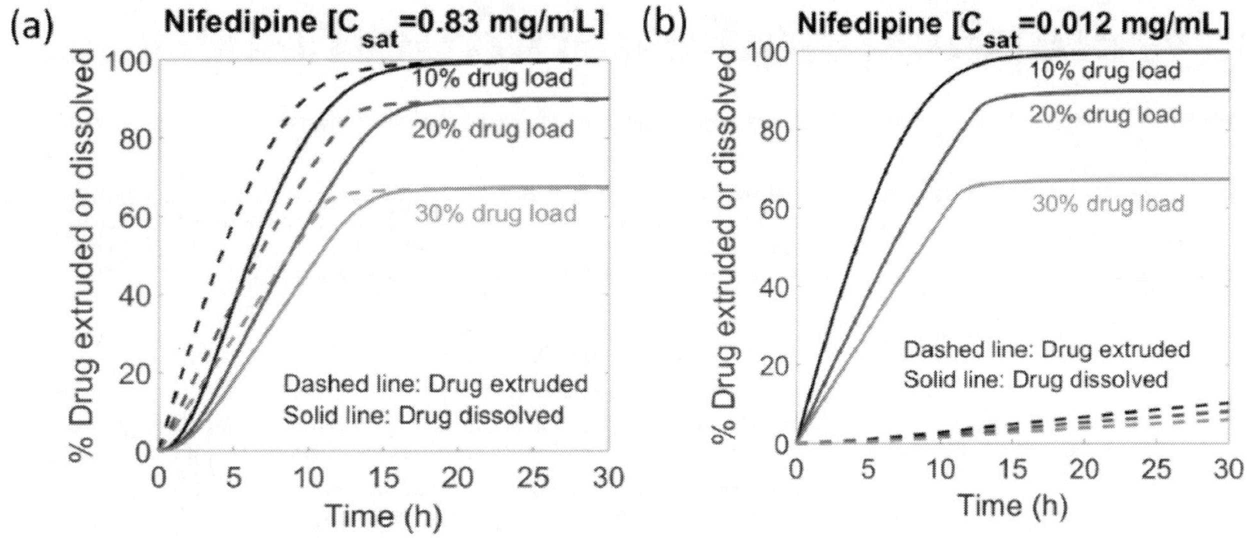

(a)
Nifedipine [C_{sat}=0.83 mg/mL]
10% drug load
20% drug load
30% drug load
Dashed line: Drug extruded
Solid line: Drug dissolved
% Drug extruded or dissolved
Time (h)
(b)
Nifedipine [C_{sat}=0.012 mg/mL]
10% drug load
20% drug load
30% drug load
Dashed line: Drug extruded
Solid line: Drug dissolved
% Drug extruded or dissolved
Time (h)

CLN3 is required for the clearance of glycerophosphodiesters from lysosomes

https://doi.org/10.1038/s41586-022-05221-y

Received: 25 April 2020

Accepted: 10 August 2022

Published online: 21 September 2022

Check for updates

Nouf N. Laqtom[1,2,3], Wentao Dong[1,2,3], Uche N. Medoh[1,2,3,4], Andrew L. Cangelosi[5,6,7], Vimisha Dharamdasani[7], Sze Ham Chan[7,11], Tenzin Kunchok[7], Caroline A. Lewis[7], Ivonne Heinze[8], Rachel Tang[9], Christian Grimm[9], An N. Dang Do[10], Forbes D. Porter[10], Alessandro Ori[8], David M. Sabatini[12,13] & Monther Abu-Remaileh[1,2,3,12] ✉

Lysosomes have many roles, including degrading macromolecules and signalling to the nucleus[1]. Lysosomal dysfunction occurs in various human conditions, such as common neurodegenerative diseases and monogenic lysosomal storage disorders (LSDs)[2–4]. For most LSDs, the causal genes have been identified but, in some, the function of the implicated gene is unknown, in part because lysosomes occupy a small fraction of the cellular volume so that changes in lysosomal contents are difficult to detect. Here we develop the LysoTag mouse for the tissue-specific isolation of intact lysosomes that are compatible with the multimodal profiling of their contents. We used the LysoTag mouse to study CLN3, a lysosomal transmembrane protein with an unknown function. In children, the loss of *CLN3* causes juvenile neuronal ceroid lipofuscinosis (Batten disease), a lethal neurodegenerative LSD. Untargeted metabolite profiling of lysosomes from the brains of mice lacking CLN3 revealed a massive accumulation of glycerophosphodiesters (GPDs)—the end products of glycerophospholipid catabolism. GPDs also accumulate in the lysosomes of CLN3-deficient cultured cells and we show that CLN3 is required for their lysosomal egress. Loss of CLN3 also disrupts glycerophospholipid catabolism in the lysosome. Finally, we found elevated levels of glycerophosphoinositol in the cerebrospinal fluid of patients with Batten disease, suggesting the potential use of glycerophosphoinositol as a disease biomarker. Our results show that CLN3 is required for the lysosomal clearance of GPDs and reveal Batten disease as a neurodegenerative LSD with a defect in glycerophospholipid metabolism.

Lysosomes are membrane-bound organelles that degrade macromolecules and clear damaged organelles as well as regulate cellular signalling, such as the activation of the mTORC1 pathway by nutrients[1,5,6]. Mutations in genes encoding lysosomal proteins cause severe disorders that are collectively known as LSDs[2,3,7–9]. Many LSDs are associated with neurodegeneration, indicating a critical role for the lysosome in neuronal homeostasis[3,4]. Consistent with this, human genetic studies have implicated impaired lysosomal function in age-associated neurodegenerative diseases, including Alzheimer disease, Parkinson's disease and frontotemporal dementia[2,10,11]. The function of several lysosomal proteins implicated in disease is incompletely understood, in part because lysosomes usually occupy only 1–3% of the cellular volume and changes in lysosomal contents can therefore be difficult to ascertain in whole cells or tissues[12].

The development of the LysoTag mouse

Given this and the increasing interest in the roles of lysosomes in physiology and disease, we adapted our LysoIP method for the rapid purification of lysosomes from cultured cells[12] for use in mouse tissues. We generated a transgenic mouse line by integrating the cDNA encoding the lysosomally localized TMEM192–3×HA fusion protein into the *Rosa26* locus downstream of a lox-stop-lox (LSL) cassette[13] (Fig. 1a). By crossing these mice with those expressing a Cre recombinase driven by the CMV promoter, we generated LysoTag mice that constitutively express TMEM192–3×HA across tissues (Fig. 1a and Extended Data Fig. 1a). TMEM192–3×HA marks lysosomes with a triple HA epitope, enabling their immunoisolation with anti-HA antibodies. Immunofluorescence-based analyses of liver and brain sections, two

[1]Department of Chemical Engineering, Stanford University, Stanford, CA, USA. [2]Department of Genetics, Stanford University, Stanford, CA, USA. [3]The Institute for Chemistry, Engineering & Medicine for Human Health (ChEM-H), Stanford University, Stanford, CA, USA. [4]Department of Biochemistry, Stanford University School of Medicine, Stanford, CA, USA. [5]Department of Biology, Massachusetts Institute of Technology, Cambridge, MA, USA. [6]Koch Institute for Integrative Cancer Research, Cambridge, MA, USA. [7]Whitehead Institute for Biomedical Research, Cambridge, MA, USA. [8]Leibniz Institute on Aging—Fritz Lipmann Institute, Jena, Germany. [9]Walther Straub Institute of Pharmacology and Toxicology, Faculty of Medicine, Ludwig-Maximilians-Universität, Munich, Germany. [10]Division of Translational Medicine, Eunice Kennedy Shriver National Institute of Child Health and Human Development, NIH, Bethesda, MD, USA. [11]Present address: Department of Pharmacology, University of Virginia School of Medicine, Charlottesville, VA, USA. [12]These authors contributed equally: David M. Sabatini, Monther Abu-Remaileh. [13]Unaffiliated: David M. Sabatini. ✉e-mail: monther@stanford.edu

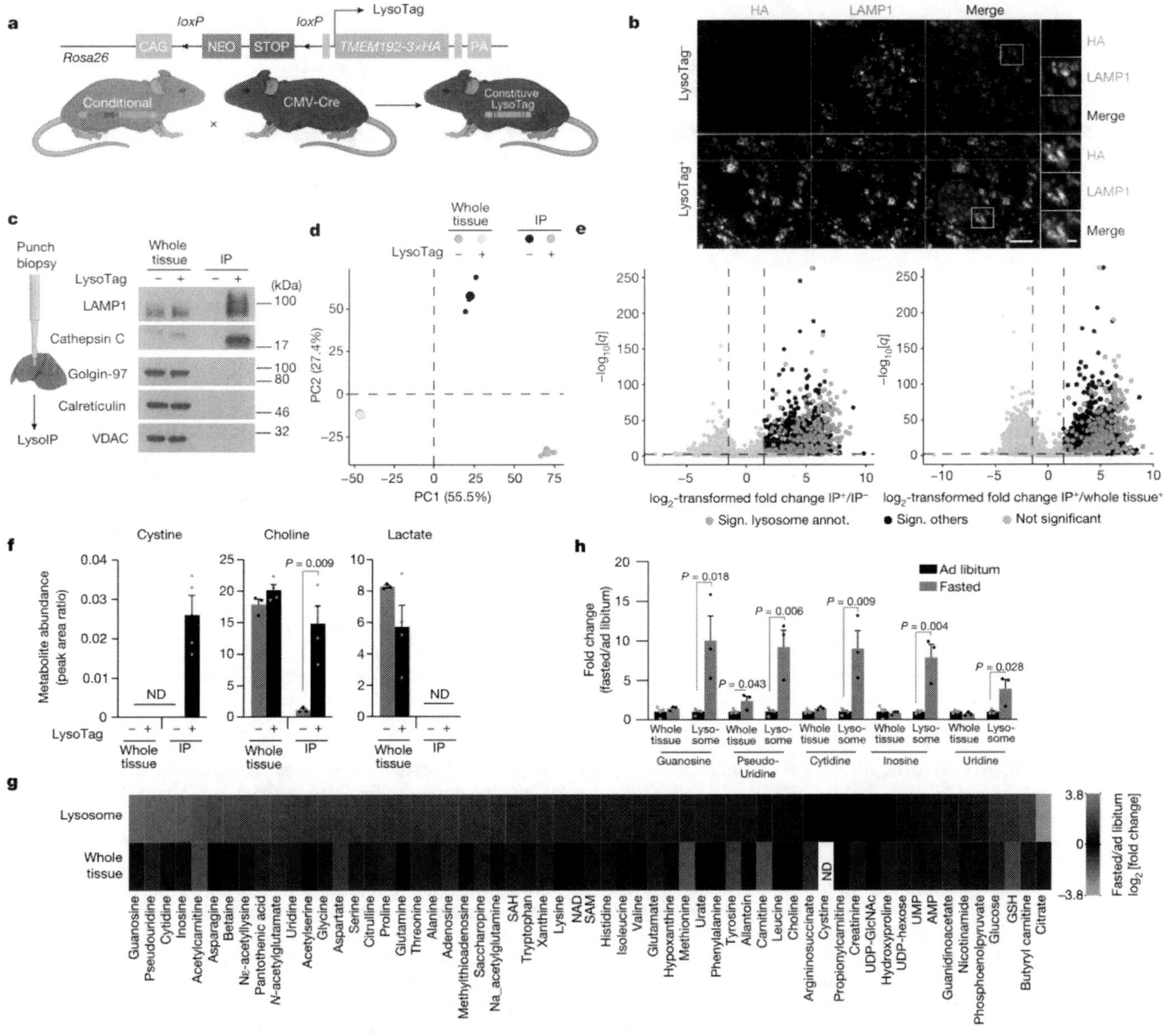

Fig. 1 | LysoTag mouse for proteomic and metabolite profiling of tissue lysosomes. **a**, Schematic of the *TMEM192-3×HA* fusion gene (LysoTag) in the *Rosa26* locus and the generation of the constitutive LysoTag line. **b**, Immunofluorescence analyses of TMEM192–3×HA (HA, green) and lysosomes (LAMP1, red) in the liver. Nuclei are shown in blue. Scale bars, 5 μm (main images) and 1 μm (insets). Micrographs are representative of three independent experiments. **c**, Immunoblot analyses of protein markers of subcellular compartments in whole tissues, purified lysosomes (immunoprecipitate: IP LysoTag⁺), and control immunoprecipitates (IP LysoTag⁻). Golgin-97, VDAC and calreticulin were used as markers for the Golgi, mitochondria and endoplasmic reticulum, respectively, and LAMP1 and cathepsin C for lysosomes. Immunoblots are representative of three independent experiments. **d**,**e**, Proteomic analyses of liver LysoIPs. **d**, PCA comparing the levels of the detected proteins. **e**, Volcano plot showing that previously annotated lysosomal proteins (orange dots) are enriched in lysosomes. Enriched unannotated proteins are shown in black. The vertical dashed lines indicate a $\log_2$-transformed fold change of >1.5 in LysoTag⁺ immunoprecipitates either over control immunoprecipitates (left) or whole tissue (right). The horizontal dashed lines indicate $q < 0.01$. $n = 3$ male mice with an average age of 6.5 weeks. Statistical analysis was performed using paired *t*-tests, performed at the peptide precursor level; *P* values were corrected for multiple testing according to ref. [42]. Annot., annotation; sign., significant. **f**, The relative abundance of the indicated metabolites in whole tissue and immunoprecipitates. ND, not detected. Data are mean ± s.e.m. $n = 4$ (LysoTag⁺) and $n = 3$ (LysoTag⁻). **g**, Heat map of the mean of $\log_2$-transformed fold changes in metabolite levels after a 16 h fast (Supplementary Table 2). **h**, Fold changes in the levels of nucleosides after fasting. For **g** and **h**, data are mean and for **h** ± s.e.m. of $n = 4$ (ad libitum) or $n = 3$ (fasted). For **f** and **h**, statistical analysis was performed using two-tailed unpaired *t*-tests. For **f**–**h**, male mice with an average age of 5.5 weeks were used. Gel source data are provided in Supplementary Fig. 1. The drawings in the schematics in **a** and **c** were created using Biorender. GlcNAc, N-acetylglucosamine; GSH, Glutathione; SAH, S-adenosylhomocysteine; SAM, S-adenosylmethionine.

tissues examined in this study, confirmed that TMEM192–3×HA colocalizes with the lysosomal marker LAMP1 (Fig. 1b and Extended Data Fig. 2b). Other than a 10–15% decrease in body weight, the constitutive LysoTag mice appear indistinguishable from their wild-type (WT) counterparts (Extended Data Fig. 1b). Importantly, comprehensive characterizations confirm that LysoTag expression has no effect on

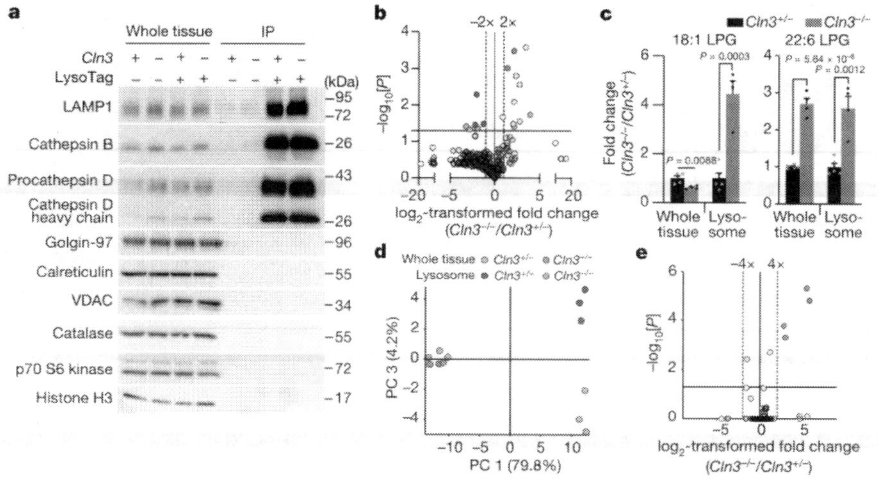

Fig. 2 | Loss of CLN3 leads to significant alterations in metabolite levels in brain lysosomes. a, Immunoblot analyses of organellar protein markers in whole-tissue fractions, purified lysosomes and control immunoprecipitates. LAMP1 and cathepsin B and D were used as markers for lysosomes. Golgin-97, calreticulin, VDAC, catalase, S6 kinase and histone H3 were used as markers for the Golgi, endoplasmic reticulum, mitochondria, peroxisomes, the cytosol and the nucleus, respectively. Immunoblots are representative of three independent experiments. **b**, Volcano plot comparing untargeted LysoIP lipidomic data from lysosomes derived from brains of *Cln3*$^{+/-}$ and *Cln3*$^{-/-}$ mice. Significantly changing lipids belong to phospholipids (yellow), LPG (red), lysophosphatidylcholine (orange) and BMP (blue). BMP/PG annotation was used when MS/MS fragmentation was not acquired (cyan). Data are provided in Supplementary Table 3. The horizontal line indicates $P = 0.05$, and the vertical dashed lines indicate a fold change of 2. *P* values were calculated using two-tailed unpaired *t*-tests. **c**, Targeted analyses of LPGs. Fold changes in LPGs in lysosomes from *Cln3*$^{+/-}$ and *Cln3*$^{-/-}$ mice were calculated after subtracting the background from control immunoprecipitates. Data are mean ± s.e.m. Statistical analysis was performed using two-tailed unpaired *t*-tests. For **b** and **c**, $n = 5$ (*Cln3*$^{+/-}$) and $n = 4$ (*Cln3*$^{-/-}$) female mice (aged 7 months). **d**,**e**, Untargeted polar metabolite analyses of brain lysosomes and tissues after CLN3 loss (Methods and Supplementary Table 4). **d**, PCA comparing the levels of the 189 unique compounds (see the main text) between whole-tissue or lysosome samples from the brains of *Cln3*$^{+/-}$ and *Cln3*$^{-/-}$ mice. **e**, The lysosomal changes in these compounds. The horizontal line indicates $P = 0.05$, and the vertical dashed lines represent fold changes of 4. $n = 3$ male mice (aged 7 months) per genotype. *P* values were calculated using analysis of variance (ANOVA) with Tukey HSD test; *P* values were corrected using the Benjamini–Hochberg method; FDR = 5%. Gel source data are provided in Supplementary Fig. 1.

lysosomal proteins, enzymatic activity or lysosomal ultrastructure (Extended Data Fig. 1a,c–f).

We modified our protocol[12] for preparing anti-HA immunoprecipitates from cultured cells for use with tissues and applied it to livers from control or LysoTag mice. Immunoprecipitates from LysoTag mice (hereafter, LysoIPs), but not from control mice, were substantially enriched for the lysosomal membrane and lumenal proteins LAMP1 and cathepsin C, respectively, but free of markers of other organelles, including the Golgi, endoplasmic reticulum and mitochondria (Fig. 1c). Furthermore, data-independent acquisition-based proteomics confirmed that the LysoIPs (Fig. 1d,e and Supplementary Table 1) were highly enriched for proteins that were previously annotated as lysosomal regardless of whether we determined lysosomal enrichment relative to the abundance of these proteins in control immunoprecipitates or whole-liver lysates (Fig. 1e). The LysoIPs were mostly depleted for proteins from other compartments, except for a select set of proteins that are mostly associated with the Golgi, suggesting a possible interaction between the organelles (Supplementary Table 1). Lysosomal membrane and lumenal proteins were enriched to similar degrees, indicating that lysosomes purified by the LysoIP protocol are intact (Supplementary Table 1). Finally, we observed enrichment of some endosomal markers, although to a lower extent than those for lysosomes (Supplementary Table 1 and Extended Data Fig. 1g).

To test the feasibility of our method for studying lysosomal metabolites, we measured cystine, which is known to be enriched in lysosomes[12,14]. Indeed, the liver LysoIPs had much higher cystine levels compared with the control immunoprecipitates or whole-liver samples, in which cystine levels were below the limit of detection (Fig. 1f). The LysoIPs also contained metabolites that we previously detected in lysosomes from cultured cells[12], including choline (Fig. 1f and Supplementary Table 2). Importantly, abundant cytosolic metabolites, such as lactate, were either not detected in the LysoIPs or equally abundant in the control immunoprecipitates that were used to reveal metabolites that bind non-specifically to the magnetic beads (Fig. 1f and Supplementary Table 2).

To test how lysosomal contents change under different physiological states, we generated targeted metabolite profiles of lysosomes from the livers of mice fed ad libitum or fasted. Whereas some of the fasting-induced changes in lysosomal metabolites were also detected at the whole-tissue level (such as acetyl-carnitine), others were observed only in lysosomes (Fig. 1g and Supplementary Table 2). For example, fasting increased the levels in lysosomes of all nucleosides, without affecting those in whole-liver samples, except for pseudouridine (Fig. 1g,h and Supplementary Table 2). Collectively, these data show that the LysoIP can be used to rapidly isolate intact and highly pure lysosomes from mouse organs and to study metabolite changes that are not detectable using traditional tissue-based metabolite profiling.

CLN3 loss alters lysosomal metabolome

To determine the ability of the LysoIP approach for deorphaning gene function, we turned to *CLN3*, the gene that is mutated in the lethal neurodegenerative disorder Batten disease[15,16]. The function of *CLN3* remains unclear[17,18]. The *CLN3* gene product is a multi-pass transmembrane protein that localizes to the lysosome[19–23] and its loss is associated with defects in multiple cellular functions, including autophagy, membrane fusion, vesicular trafficking and arginine transport[17,18]. Although its localization might suggest that it transports molecules across the lysosomal membrane, only recently has CLN3 been suggested to be an atypical solute carrier based on structural predictions[24].

To identify changes in the lysosomal metabolome caused by CLN3 loss, we crossed the LysoTag mice with *Cln3*$^{-/-}$ mice[25,26] (Extended Data Fig. 2a). As in the liver, TMEM192–3×HA colocalized with LAMP1-positive lysosomes in the brains of all genotypes (Extended Data Fig. 2b). Consistent

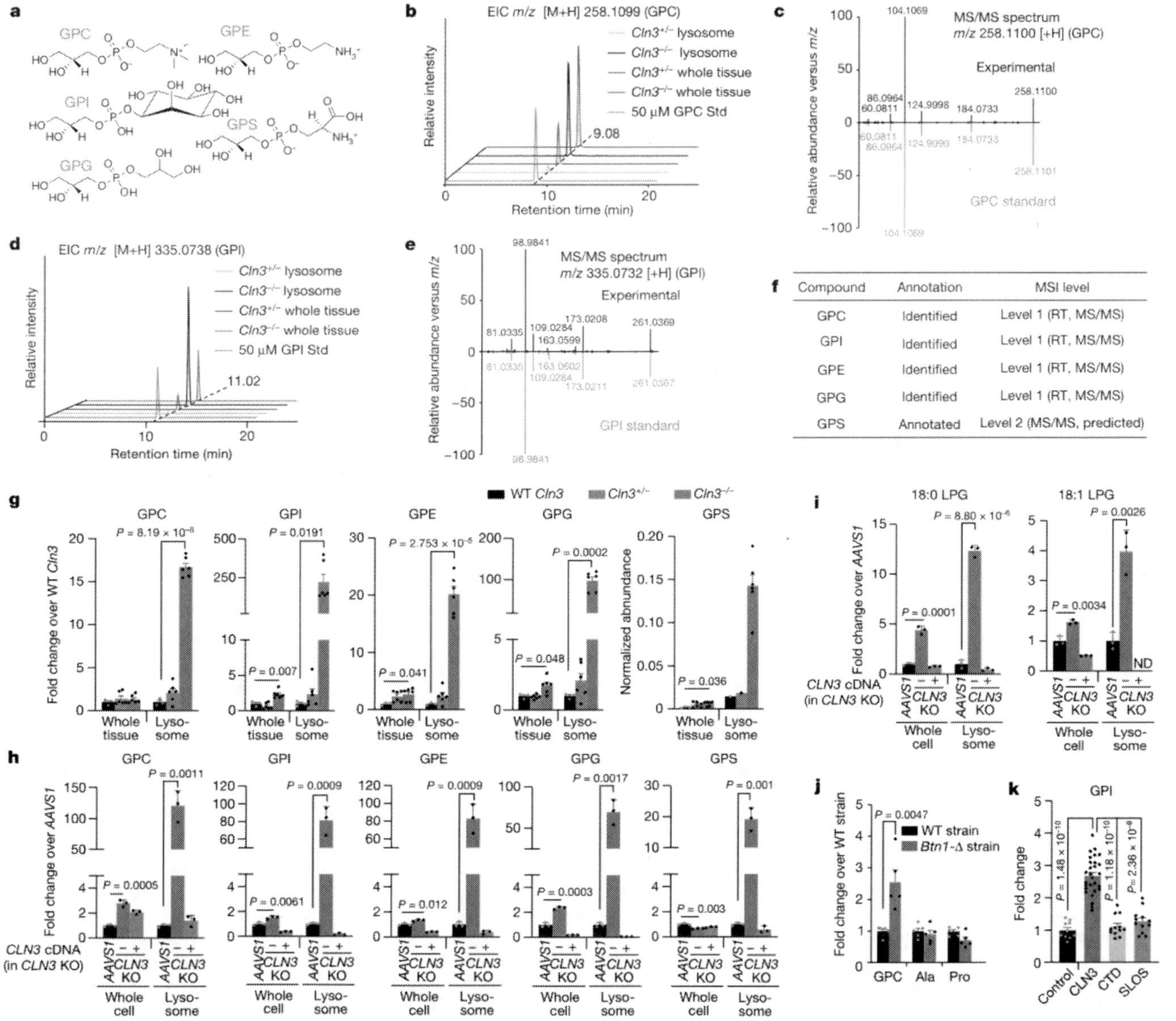

Fig. 3 | GPDs accumulate in lysosomes after CLN3 loss. a, The chemical structure of GPDs in this study. **b**, Extracted ion chromatogram (EIC) analysis of GPC across a range of samples alongside an authentic standard (Std) showing a matching retention time. **c**, Mirror plot for an MS/MS spectrum of GPC in a representative sample (lysosomes from *Cln3*$^{-/-}$ mouse brain) and an authentic standard. **d**, Extracted ion chromatogram as described in **b** but for GPI. **e**, Mirror plot for an MS/MS spectrum as described in **c** but for GPI. **f**, The level of confidence in the identification of GPDs in this study according to MSI guidelines. **g**, Targeted analyses of GPDs in brain lysosomes. Fold changes between lysosomes were calculated after subtracting the background signal in control immunoprecipitates and normalizing to methionine. Glycerophosphoserine (GPS) was detected in only one WT and *Cln3*$^{+/-}$ lysosomal sample; thus, its abundance was reported instead. Data are mean ± s.e.m. $n = 3$ (WT *Cln3*), $n = 6$ (*Cln3*$^{+/-}$) and $n = 6$ (*Cln3*$^{-/-}$). *P* values presented are between WT *Cln3* and *Cln3*$^{-/-}$. Male mice from two independent experiments, including that in Fig. 2e. **h**, Fold change in the levels of GPDs between HEK293T cells and their lysosomes after *CLN3* knockout and rescue. sgRNA targeting the *AAVS1* locus was used as a control. **i**, Targeted analyses of LPGs from cells as described in **h**. For **h** and **i**, data are mean ± s.d. of $n = 3$ biologically independent samples. **j**, The fold change in the levels of GPC, alanine (Ala) and proline (Pro) between WT and *Btn1-Δ* yeast strains. Data are mean ± s.e.m. $n = 5$ biologically independent samples. **k**, The fold change in GPI levels in the cerebrospinal fluid from patients with CLN3 Batten disease (CLN3, $n = 28$), creatine transporter deficiency (CTD, $n = 12$) and Smith–Lemli–Opitz Syndrome (SLOS, $n = 12$) compared with anonymized paediatric control individuals (control, $n = 10$). Data are mean ± s.e.m. Characteristics of the study participants are provided in Supplementary Table 5. For **g**–**k**, statistical analysis was performed using two-tailed unpaired *t*-tests.

with previous reports, LAMP1-positive compartments in neurons lacking CLN3 were noticeably more abundant than in control cells[27] (Extended Data Fig. 2b). Moreover, our protocol successfully yielded pure and intact lysosomes from the brain, regardless of CLN3 status (Fig. 2a).

As alterations in lysosome-mediated lipid catabolism have been associated with multiple forms of LSDs[28,29], we used untargeted lipidomics (Methods) to analyse brain lysosomes from the mice lacking CLN3. Consistent with previous reports[30], we detected a significant decrease in tissue and lysosomal levels of bis(monoacylglycero)phosphates (BMPs), a class of lysosomal lipids (Fig. 2b, Extended Data Fig. 2c and Supplementary Table 3). Interestingly, we also found significant increases in various glycerophospholipids (hereafter phospholipids) (Supplementary Table 3), including the degradation intermediate lysophosphatidylglycerol (LPG) (Fig. 2b,c and Extended Data Fig. 2c). Although these changes

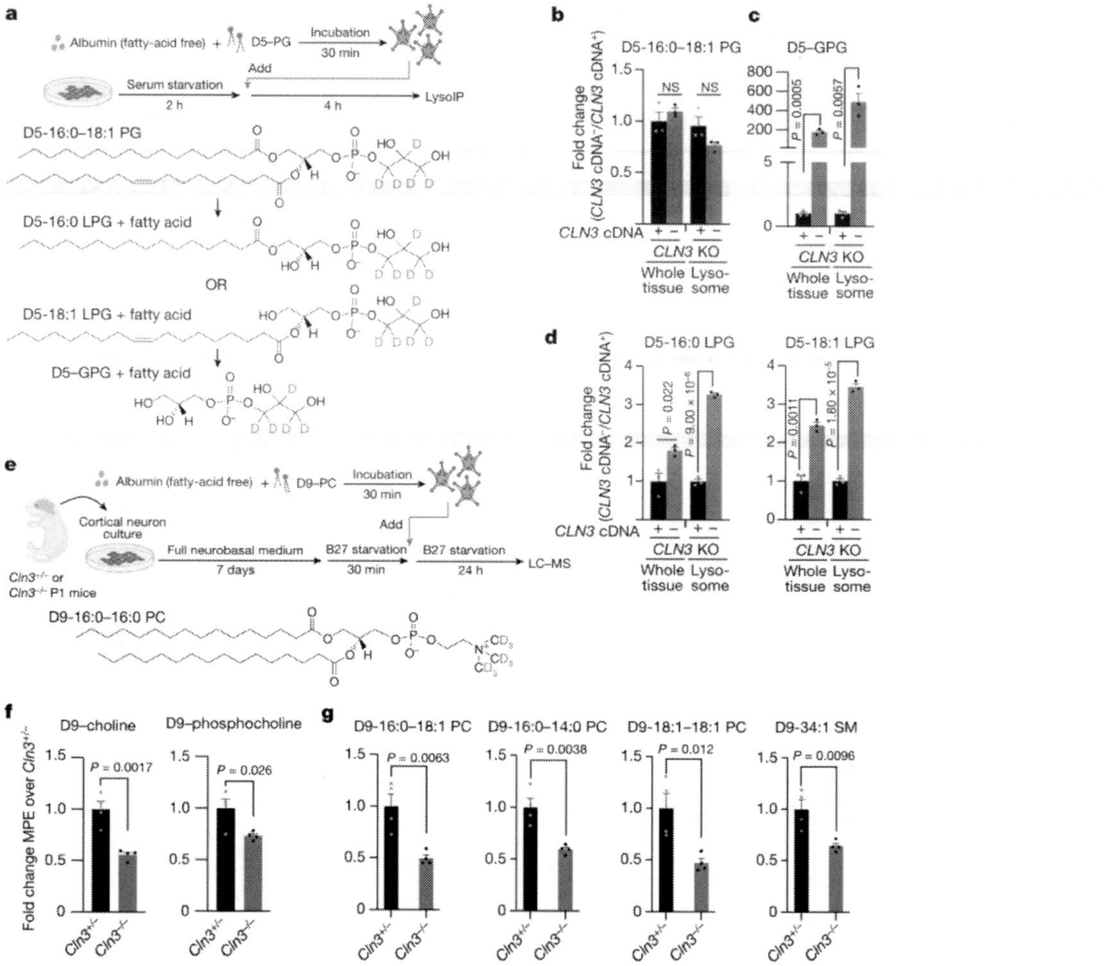

Fig. 4 | CLN3 is required for the efflux of GPDs out of lysosomes. **a**, An approach to deliver exogenously added deuterated phospholipids to lysosomes. The cell culture drawing in the schematic was created using Biorender. **b**, CLN3 loss does not affect the cellular uptake of exogenous D5-PG (deuterated phosphatidylglycerol D5-16:0–18:1 PG). Fold changes in the whole-cell and lysosomal abundances of D5-PG in *CLN3*-KO cells relative to those in the *CLN3*-KO cells rescued with the *CLN3* cDNA. **c**, Loss of CLN3 blocks the efflux of GPG from lysosomes. Data are presented as described in **b** but for D5-GPG. **d**, LPGs generated in lysosomes accumulate upon CLN3 loss. Data are presented as in **b** but for the two possible D5-LPGs that are generated from the degradation of the exogenously added D5-PG. For **b**–**d**, data are mean ± s.d. of $n = 3$. **e**–**g**, GPC-derived choline contributes to lipid biosynthesis in a CLN3-dependent manner in primary cortical neurons. **e**, The approach to trace the metabolic fate of the choline moiety of GPC, which is generated from phosphatidylcholine (PC) degradation in the lysosome. PC with nine deuterium atoms in its choline headgroup (D9-16:0–16:0 PC) was delivered to lysosomes in primary cultured neurons isolated from the cerebral cortex of *Cln3*$^{+/-}$ (control) or *Cln3*$^{-/-}$ mice. The mouse and cell culture drawings are from BioRender. **f**, CLN3 loss significantly reduces the cellular levels of deuterated phosphocholine and its precursor choline. **g**, CLN3 loss leads to a reduced contribution of choline derived from lysosomal GPC to the biosynthesis of PC and sphingomyelin (SM) in the cell. Data are presented as fold changes in the whole-cell molar percent enrichment (MPE) of D9–choline-containing metabolites in cortical neuron cultures prepared from *Cln3*$^{-/-}$ mice relative to those from *Cln3*$^{+/-}$ animals. For **f** and **g**, data are mean ± s.e.m. of $n = 4$. For **b**–**d**, **f** and **g**, statistical analysis was performed using two-tailed unpaired *t*-tests; and *n* values represents biologically independent samples. NS, not significant.

could represent a link between CLN3 function and phospholipid catabolism in the lysosome, which was previously noted in yeast models[31], it is not clear how they are directly linked to CLN3 loss.

Thus, we turned to untargeted polar metabolomics to search for additional lysosomal metabolites that could be affected by CLN3 loss (Methods and Extended Data Fig. 3a). We detected 1,073 putative compounds in lysosomes after background removal and signal thresholding (Supplementary Table 4). After further filtering for compounds in known compound databases and with reliable fragmentation data, we arrived at 189 unique compounds for which we had various degrees of confidence in their identities based on the Metabolomics Standards Initiative (MSI) guidelines (Extended Data Fig. 3a and Supplementary Table 4). Principal component analyses (PCA) of the 189 compounds revealed that, although the metabolic landscapes of the *Cln3*$^{+/-}$ and *Cln3*$^{-/-}$ whole-brain samples were similar, those in their respective lysosomes were distinct (Fig. 2d); four compounds at high levels in lysosomes lacking CLN3 were the most differentially abundant (Fig. 2e). Critically, these same four compounds were the most significantly affected by CLN3 loss independently of how we filtered the datasets (Supplementary Table 4), indicating that CLN3 function affects a specific set of metabolites in lysosomes rather than the whole lysosomal metabolome.

CLN3-deficient lysosomes accumulate GPDs

To determine the chemical identity of the accumulating metabolites, we used MS1 as well as tandem mass spectrometry (MS/MS) fragmentation data (Extended Data Fig. 3a). This effort indicated that the four metabolites were probably all GPDs, which are formed by the deacylation of phospholipids (Fig. 3a and Supplementary Table 4). We tentatively annotated the compounds as glycerophosphoglycerol (GPG), glycerophosphocholine (GPC), glycerophosphoinositol (GPI) and glycerophosphoethanolamine (GPE), which represent four of the five major GPDs (Fig. 3a). To validate the compounds as GPDs, we showed that commercially available standards for GPC (Fig. 3b,c and Extended Data Fig. 3b) and GPI (Fig. 3d,e and Extended Data Fig. 3c) had retention times and MS/MS spectra matching those in biological samples. For GPE and GPG, owing to the lack of commercial standards, we generated these metabolites in house, and validated that the retention times, *m/z* values and MS/MS spectrum from the biological samples matched those of the standards (Extended Data Fig. 3d–f) and, when available, previously experimentally reported fragments[32]. Although the fifth major GPD, glycerophosphoserine (GPS), was not detected in the original untargeted metabolite profiles, follow-up work showed that its lysosomal levels are also increased after CLN3 loss (Fig. 3g). Its identity was validated by comparing *m/z* values and MS/MS spectrum in our samples with those of in silico fragmentation patterns generated using competitive fragmentation modelling-ID (CFM-ID)[33] (Extended Data Fig. 3g). Together, our approach rigorously identified and annotated the metabolites that accumulate in the lysosome after CLN3 loss according to MSI guidelines[34] (Fig. 3f). We also manually extracted ion chromatograms for GPDs, which confirmed large and significant accumulations of the GPDs in brain lysosomes from the Batten disease model, but also showed almost no or only very modest increases in their corresponding levels in the whole-brain samples (Fig. 3g). These data are consistent with lysosomes occupying a small fraction of the total cellular volume and underscore the value of directly measuring metabolites in lysosomes.

To rule out that the increase in GPDs in the brain lysosomes of the Batten disease mice is a consequence of neurodevelopment defects, we generated and analysed HEK293T cells (Extended Data Fig. 3h) that are null for *CLN3* and also express TMEM192–3×HA. In lysosomes from these cells, all five GPDs also accumulated to very high levels and reverted to control amounts or below after expression of the *CLN3* cDNA (Fig. 3h and Extended Data Fig. 3i). Importantly, we found that, regardless of the tagging system used to purify lysosomes, similar levels of GPD accumulation were observed after CLN3 loss (Extended Data Fig. 3j). As in the brain, the HEK293T lysosomes lacking CLN3 also had increases in LPG levels (Fig. 3i).

Human CLN3 can complement the functions of its yeast homologue Btn1p in multiple lysosomal assays[17], indicating that they share the same function. Indeed, yeast lacking Btn1p significantly accumulated GPC, the only detectable GPD in yeast, suggesting that the function of CLN3 in GPD metabolism is conserved (Fig. 3j). Importantly, we found elevated levels of GPI in the cerebrospinal fluid (CSF) of individuals with CLN3 Batten disease compared to that from individuals without CLN3 disease[35] (Fig. 3k), suggesting its potential use as a clinical biomarker.

Thus, the use of untargeted metabolomics in combination with the LysoIP approach reveals that, in the mouse brain and in human cells in culture, the loss of CLN3 causes a large lysosomal accumulation of GPDs along with a more modest increase in lysophospholipids, which are the metabolites directly upstream of GPDs in the pathway for phospholipid catabolism. Although the specific lysosomal phospholipases that generate GPDs from phospholipids have not been defined, we note that lysosomes do contain the phospholipase A/B activities required for their release from phospholipids[36].

CLN3 is required for the efflux of GPDs

The acidic lysosomal lumen establishes a proton gradient with the cytosol that drives the efflux of several metabolites, including GPDs[12] (Extended Data Fig. 4c,d). A plausible explanation for GPD accumulation is that CLN3 loss increases the lysosomal pH, which attenuates GPD egress from the lysosome. However, consistent with previous reports[37], CLN3 loss in HEK293T cells does not increase lysosomal pH and, if anything, leads to a small decrease (Extended Data Fig. 4a,b). Consistent with this, after CLN3 loss, we observed no lysosomal accumulation of amino acids (proline, alanine and glutamate) of which the efflux from the lysosome is dependent on pH—functionally confirming our pH measurements (Extended Data Fig. 4d). Thus, the lysosomal accumulation of GPDs in CLN3-deficient lysosomes is independent of perturbations in lysosomal pH.

Given these data and that CLN3 is a multi-pass transmembrane protein, we hypothesized that CLN3 is either a lysosomal glycerophosphodiesterase (GDE) required for hydrolysing GPDs into glycerol 3-phosphate and their corresponding headgroups—an activity that is mostly associated with membrane proteins[38]—or, alternatively, required for the efflux from lysosomes of GPDs produced by the lysosomal catabolism of phospholipids. To test the former, we purified CLN3 and, as a positive control, the established glycerophosphodiesterase GDE1. Although GDE1 readily hydrolysed deuterated GPC in vitro, we found no such activity for CLN3 (Extended Data Fig. 4e). These results are consistent with previous reports indicating that GPDs are the final products of phospholipid degradation in mammalian lysosomes[36].

To test the possibility that CLN3 is required for the lysosomal efflux of GPDs, we used an assay in HEK293T cells in which, through endocytosis, we delivered to lysosomes a phospholipid with a labelled head group that, after lysosomal catabolism, produces a labelled GPD. We monitored the fate of the labelled GPD using mass spectrometry (Fig. 4a). Specifically, we conjugated phosphatidylglycerol with a 5×deuterium-labelled head group (D5-PG) to bovine serum albumin (BSA) and added it to the medium of serum-starved HEK293T cells expressing TMEM192–3×HA (Fig. 4a). The relative amounts of the deuterated phosphatidylglycerol were the same in the whole-cell and lysosomal fractions of control and *CLN3*-null cells, indicating that loss of CLN3 does not affect the uptake of the phospholipid into lysosomes (Fig. 4b). Notably, we could detect only a small signal for deuterated GPG—the GPD released from the labelled phosphatidylglycerol—in the whole-cell and lysosomal samples from the control cells, consistent with GPG being quickly effluxed from normal lysosomes and further catabolized in the cytosol. By contrast, in the *CLN3*-null cells, deuterated GPG readily accumulated in lysosomes and was also easily detected in whole-cell samples (Fig. 4c). These results indicate that, in cells lacking CLN3, the labelled GPG is trapped in lysosomes and is therefore not catabolized in the cytosol (Fig. 4c). Consistent with our previous metabolite profiles (Fig 3i), in the *CLN3*-null cells, we also observed a significant accumulation of labelled LPGs (Fig. 4d)—the metabolites directly upstream of GPD. Thus, we conclude that CLN3 is required for transporting the GPDs released from phospholipids out of the lysosome.

To further validate this model and to understand the effect of impaired GPD recycling in neurons, we used phosphatidylcholine (PC) with a deuterium-labelled head group (D9-PC) to trace the metabolic fate of its choline moiety in the cell, which is generated by the catabolism of GPC by GDE activity in the cytoplasm (Fig. 4e). CLN3 loss significantly reduced the pools of labelled choline and phosphocholine in primary neuronal cultures (Fig. 4f). As a consequence, we observed a significant reduction in the deuterated isotopomers of multiple choline-containing lipid species, including those of PC and sphingomyelin (Fig. 4g). Importantly, loss of CLN3 did not decrease the uptake of the tracer into neuronal cells (Extended Data Fig. 4f), and did not affect lipid biosynthesis or turnover as indicated by the similar labelling patterns, regardless of CLN3 status, when free D9-choline was used as a tracer (Extended Data Fig. 4g). Together, these data further support a critical role for CLN3 in the egress of GPDs from the lysosome for their catabolism and use by the rest of the cell.

Discussion

Interestingly, over 60 years ago Christian deDuve discovered that GPDs are terminal products of phospholipid degradation in the lysosome[36]. This finding implies that a mechanism must exist to efflux GPDs from the lysosome so that they can be degraded by GDEs in the cytosol[36,38]. However, no transporter for GPDs has been identified in vertebrates, although a plasma membrane GPD transporter has been reported in yeast[39]. Here we show that CLN3, the loss of which causes a severe neurodegenerative disease in children, is required for the efflux of GPDs from the lysosome. The membrane localization and predicted atypical solute carrier structure[24] of CLN3 suggests that it is a lysosomal GPD exporter or a key component of one, although further research using purified CLN3 will be needed to test this possibility. Moreover, the substantial accumulation of GPDs caused by CLN3 loss might inhibit as yet unidentified enzymes upstream of GPD production to cause the increase that we observed in lysophospholipids, which are known to be neurotoxic[40]. Indeed, GPDs have previously been reported to inhibit lysophospholipase activities in vitro[41]. Furthermore, our tracing experiments suggest that CLN3 loss might contribute to neurodegeneration by causing the extralysosomal deprivation of nutrients that are essential for brain cell function and membrane integrity such as choline in addition to the potentially toxic effects of storage material in the lysosome[2].

Our work provides a framework for future studies into how the loss of CLN3 might affect cellular homeostasis and, more generally, for using the in vivo LysoIP method in combination with untargeted metabolite profiling for understanding the function of disease-associated lysosomal proteins.

Online content

Any methods, additional references, Nature Research reporting summaries, source data, extended data, supplementary information, acknowledgements, peer review information; details of author contributions and competing interests; and statements of data and code availability are available at https://doi.org/10.1038/s41586-022-05221-y.

1. Ballabio, A. & Bonifacino, J. S. Lysosomes as dynamic regulators of cell and organismal homeostasis. *Nat. Rev. Mol. Cell Biol.* **21**, 101–118 (2020).
2. Medoh, U. N., Chen, J. Y. & Abu-Remaileh, M. Lessons from metabolic perturbations in lysosomal storage disorders for neurodegeneration. *Curr. Opin. Syst. Biol.* **29**, 100408 (2022).
3. Platt, F. M., d'Azzo, A., Davidson, B. L., Neufeld, E. F. & Tifft, C. J. Lysosomal storage diseases. *Nat. Rev. Dis. Primers* **4**, 27 (2018).
4. Ferguson, S. M. Neuronal lysosomes. *Neurosci. Lett.* **697**, 1–9 (2019).
5. Perera, R. M. & Zoncu, R. The lysosome as a regulatory hub. *Annu. Rev. Cell Dev. Biol.* **32**, 223–253 (2016).
6. Savini, M., Zhao, Q. & Wang, M. C. Lysosomes: signaling hubs for metabolic sensing and longevity. *Trends Cell Biol.* **29**, 876–887 (2019).
7. Ballabio, A. & Gieselmann, V. Lysosomal disorders: from storage to cellular damage. *Biochim. Biophys. Acta* **1793**, 684–696 (2009).
8. Boustany, R. M. Lysosomal storage diseases—the horizon expands. *Nat. Rev. Neurol.* **9**, 583–598 (2013).
9. Marques, A. R. A. & Saftig, P. Lysosomal storage disorders—challenges, concepts and avenues for therapy: beyond rare diseases. *J. Cell Sci.* **132**, jcs221739 (2019).
10. Wallings, R. L., Humble, S. W., Ward, M. E. & Wade-Martins, R. Lysosomal dysfunction at the centre of Parkinson's disease and frontotemporal dementia/amyotrophic lateral sclerosis. *Trends Neurosci.* **42**, 899–912 (2019).
11. Wang, C., Telpoukhovskaia, M. A., Bahr, B. A., Chen, X. & Gan, L. Endo-lysosomal dysfunction: a converging mechanism in neurodegenerative diseases. *Curr. Opin. Neurobiol.* **48**, 52–58 (2018).
12. Abu-Remaileh, M. et al. Lysosomal metabolomics reveals V-ATPase- and mTOR-dependent regulation of amino acid efflux from lysosomes. *Science* **358**, 807–813 (2017).
13. Thai, T. H. et al. Regulation of the germinal center response by microRNA-155. *Science* **316**, 604–608 (2007).
14. Pisoni, R. L., Acker, T. L., Lisowski, K. M., Lemons, R. M. & Thoene, J. G. A cysteine-specific lysosomal transport system provides a major route for the delivery of thiol to human fibroblast lysosomes: possible role in supporting lysosomal proteolysis. *J. Cell Biol.* **110**, 327–335 (1990).
15. Eiberg, H., Gardiner, R. M. & Mohr, J. Batten disease (Spielmeyer-Sjogren disease) and haptoglobins (HP): indication of linkage and assignment to chr. 16. *Clin. Genet.* **36**, 217–218 (1989).
16. Lerner, T.J. et al. Isolation of a novel gene underlying Batten disease, *CLN3*. *Cell* **82**, 949–957 (1995).
17. Mirza, M. et al. The *CLN3* gene and protein: what we know. *Mol. Genet. Genomic Med.* **7**, e859 (2019).
18. Butz, E. S., Chandrachud, U., Mole, S. E. & Cotman, S. L. Moving towards a new era of genomics in the neuronal ceroid lipofuscinoses. *Biochim. Biophys. Acta Mol. Basis Dis.* **1866**, 165571 (2019).
19. Jarvela, I. et al. Biosynthesis and intracellular targeting of the CLN3 protein defective in Batten disease. *Hum. Mol. Genet.* **7**, 85–90 (1998).
20. Storch, S., Pohl, S. & Braulke, T. A dileucine motif and a cluster of acidic amino acids in the second cytoplasmic domain of the batten disease-related CLN3 protein are required for efficient lysosomal targeting. *J. Biol. Chem.* **279**, 53625–53634 (2004).
21. Mao, Q., Foster, B. J., Xia, H. & Davidson, B. L. Membrane topology of CLN3, the protein underlying Batten disease. *FEBS Lett.* **541**, 40–46 (2003).
22. Ezaki, J. et al. Characterization of Cln3p, the gene product responsible for juvenile neuronal ceroid lipofuscinosis, as a lysosomal integral membrane glycoprotein. *J. Neurochem.* **87**, 1296–1308 (2003).
23. Oetjen, S., Kuhl, D. & Hermey, G. Revisiting the neuronal localization and trafficking of CLN3 in juvenile neuronal ceroid lipofuscinosis. *J. Neurochem.* **139**, 456–470 (2016).
24. Perland, E., Bagchi, S., Klaesson, A. & Fredriksson, R. Characteristics of 29 novel atypical solute carriers of major facilitator superfamily type: evolutionary conservation, predicted structure and neuronal co-expression. *Open Biol.* **7**, 170142 (2017).
25. Mitchison, H. M. et al. Targeted disruption of the *Cln3* gene provides a mouse model for Batten disease. The Batten Mouse Model Consortium [corrected]. *Neurobiol. Dis.* **6**, 321–334 (1999).
26. Kovacs, A. D. & Pearce, D. A. Finding the most appropriate mouse model of juvenile CLN3 (Batten) disease for therapeutic studies: the importance of genetic background and gender. *Dis. Model. Mech.* **8**, 351–361 (2015).
27. Lojewski, X. et al. Human iPSC models of neuronal ceroid lipofuscinosis capture distinct effects of TPP1 and CLN3 mutations on the endocytic pathway. *Hum. Mol. Genet.* **23**, 2005–2022 (2014).
28. Platt, F. M. Sphingolipid lysosomal storage disorders. *Nature* **510**, 68–75 (2014).
29. Fuller, M. & Futerman, A. H. The brain lipidome in neurodegenerative lysosomal storage disorders. *Biochem. Biophys. Res. Commun.* **504**, 623–628 (2018).
30. Hobert, J. A. & Dawson, G. A novel role of the Batten disease gene *CLN3*: association with BMP synthesis. *Biochem. Biophys. Res. Commun.* **358**, 111–116 (2007).
31. Padilla-Lopez, S., Langager, D., Chan, C. H. & Pearce, D. A. *BTN1*, the *Saccharomyces cerevisiae* homolog to the human Batten disease gene, is involved in phospholipid distribution. *Dis. Model. Mech.* **5**, 191–199 (2012).
32. Kopp, F. et al. The glycerophospho metabolome and its influence on amino acid homeostasis revealed by brain metabolomics of GDE1(−/−) mice. *Chem. Biol.* **17**, 831–840 (2010).
33. Allen, F., Pon, A., Wilson, M., Greiner, R. & Wishart, D. CFM-ID: a web server for annotation, spectrum prediction and metabolite identification from tandem mass spectra. *Nucleic Acids Res.* **42**, W94–W99 (2014).
34. Sumner, L. W. et al. Proposed minimum reporting standards for chemical analysis Chemical Analysis Working Group (CAWG) Metabolomics Standards Initiative (MSI). *Metabolomics* **3**, 211–221 (2007).
35. Dang Do, A. N. et al. Neurofilament light chain levels correlate with clinical measures in CLN3 disease. *Genet. Med.* **23**, 751–757 (2021).
36. Fowler, S. & De Duve, C. Digestive activity of lysosomes. 3. The digestion of lipids by extracts of rat liver lysosomes. *J. Biol. Chem.* **244**, 471–481 (1969).
37. Schmidtke, C. et al. Lysosomal proteome analysis reveals that CLN3-defective cells have multiple enzyme deficiencies associated with changes in intracellular trafficking. *J. Biol. Chem.* **294**, 9592–9604 (2019).
38. Corda, D. et al. The emerging physiological roles of the glycerophosphodiesterase family. *FEBS J.* **281**, 998–1016 (2014).
39. Patton-Vogt, J. Transport and metabolism of glycerophosphodiesters produced through phospholipid deacylation. *Biochim. Biophys. Acta* **1771**, 337–342 (2007).
40. Rigoni, M. et al. Equivalent effects of snake PLA2 neurotoxins and lysophospholipid-fatty acid mixtures. *Science* **310**, 1678–1680 (2005).
41. Fallbrook, A., Turenne, S. D., Mamalias, N., Kish, S. J. & Ross, B. M. Phosphatidylcholine and phosphatidylethanolamine metabolites may regulate brain phospholipid catabolism via inhibition of lysophospholipase activity. *Brain Res.* **834**, 207–210 (1999).
42. Storey, J. D. A direct approach to false discovery rates. *J. R. Stat. Soc. B* **64**, 479–498 (2002).

Methods

Mouse studies

Mice were maintained on a standard light–dark cycle with access to food and water ad libitum in a room with a controlled temperature (22 °C) and humidity (around 50%). Cages were cleaned every 4–5 days and supplies of water and food were checked daily. For the generation of *Rosa26;lox-stop-lox-TMEM192-3×HA*, the *TMEM192* coding sequence containing a C-terminal 3×HA epitope tag was inserted into the CTV vector (Addgene plasmid,15912)[13], linearized with AsiSI and electroporated into embryonic stem cells of the 129 background. Embryonic stem cell colonies were picked, and successfully targeted clones were identified by long-range PCR amplifying the regions spanning each homology arm. Positive embryonic stem cell clones with proper integration into the *Rosa26* locus were then injected into blastocysts and transferred into pseudopregnant female mice to obtain chimeric mice. Chimeras were bred to C57BL/6J mice for at least three generations before maintaining the line in a homozygous form. To generate a constitutive LysoTag expressing line, *Rosa26;lox-stop-lox-TMEM192-3×HA* mice were crossed with CMV-Cre transgenic mice of the C57BL/6J background, excision of lox-stop-lox was validated using our genotyping PCR method (details are provided below), and positive progeny were bred to C57BL/6J mice for at least three generations. The conditional LysoTag mouse was deposited at the Jackson Laboratory (strain, 035401) with the common name: conditional LysoTag (*Rosa26;lox-stop-lox-TMEM192-3×HA*).

The Batten disease mouse model (*Cln3*$^{-/-}$) was purchased from the Jackson Laboratory (029471, B6.129S6-Cln3^{tm1Nbm}/J) and maintained in the homozygous state. For LysoIP experiments, *Cln3*$^{-/-}$ mice were bred with LysoTag mice to generate *LysoTag;Cln3*$^{+/-}$ (used as controls) and the *LysoTag;Cln3*$^{-/-}$ experimental group. Statistical methods were not used to predetermine the sample size. Sample blinding was not performed for most of the experiments as they did not involve qualitative scoring (see the Reporting summary for details).

All of the procedures involving mice were carried out in accordance with the approved guidelines by the Institutional Animal Care and Use Committee at MIT, the Whitehead Institute for Biomedical Research and Stanford University.

Mouse genotyping

Genomic DNA was isolated from ear biopsies using QuickExtract DNA Extraction Solution (Lucigen) and PCR was performed using GoTaq DNA polymerase (Promega). To check for the presence of the *lox-stop-lox-TMEM192-3×HA* cassette in the *Rosa26* locus, two pairs of primers were designed. The non-targeted allele is amplified by primer pair P1, which produces a PCR product of 588 bp; by contrast, if the *lox-stop-lox-TMEM192-3×HA* cassette is present, it is amplified by primer pair P2, which gives a 215 bp product. The primer sequences are as follows: P1F, 5′-CAGTAAGGGAGCTGCAGTGG-3′; P1R, 5′-GCAGAAGG AGCGGGAGAAAT-3′; P2F, 5′-TACCTGAAGCGACACAATGC-3′ (within *TMEM192* sequence); P2R, 5′-CCGCTCCCTCCAGCATAATC-3′ (within HA sequence). The PCR reaction was performed with four primers in multiplex using the following conditions: 95 °C for 3 min followed by 35 cycles of 95 °C for 30 s, 56 °C for 30 s and 72 °C for 2 min, and a final extension step was added for 72 °C for 5 min. To test for the excision of the *lox-stop-lox* fragment by the Cre recombinase, two forward primers were designed; one within the *loxp* stop cassette and another upstream the first *loxp* site, both having a common reverse primer within the *TMEM192* sequence. The non-excised allele generates a 331 bp product, whereas successful excision of the *loxp-stop-loxp* site gives a 193 bp product. The primer sequences were as follows: P3F1, 5′-GGGCAACGTGCTGGTTATTG-3′; P3F2, 5′-CCTCCCCCTGAACCTGAAAC-3′; P3R, 5′GAAGCTGGGCATC CAGAAGT-3′. The PCR reaction was performed with three primers using the following conditions: 95 °C for 3 min followed by 35 cycles of 95 °C for 30 s, 60 °C for 30 s and 72 °C for 1:20 min, and a final extension step was added for 72 °C for 5 min.

Antibodies, chemicals and plasmids

Detailed information on the antibodies, chemicals and plasmids used in this study is provided in Supplementary Table 6. Additional reagents included the following secondary antibodies for immunofluorescence (used at 1:1,000): goat anti-rat Alexa Fluor 488 (Thermo Fisher Scientific, A-11006), goat anti-rabbit Alexa Fluor 488 (Thermo Fisher Scientific, A-21206), goat anti-mouse Alexa Fluor 647 (Thermo Fisher Scientific, A-21240), goat anti-mouse Alexa Fluor 488 (Thermo Fisher Scientific, A-11001), and goat anti-rabbit Alexa Fluor 594 (Thermo Fisher Scientific, A32740); the following secondary antibodies for immunoblotting (used at 1:3,000): anti-rabbit HRP (Cell Signaling Technology (CST), 7074), anti-mouse HRP (CST, 7076), anti-rat HRP (CST, 7077), and normal donkey serum, which were purchased from CST; Inactivated fetal calf serum (IFS) from Invitrogen; DMEM high glucose, anti-HA magnetic beads and ECL2 western blotting substrate from Thermo Fisher Scientific; XtremeGene9 DNA Transfection Reagent, Complete Protease Inhibitor Cocktail and PhosSTOP phosphatase inhibitor from Roche; and QuickExtract DNA Extraction Solution from Lucigen.

Rapid isolation of lysosomes from brain and liver tissues (LysoIP)

Immediately after euthanasia, mouse brains were collected and cerebral hemispheres dissected on an ice-cold plastic dish. For liver samples, we obtained a small round piece using a biopsy punch (4 mm diameter), which helped to reduce variability between samples and minimize processing time prior to the LysoIP. The liver tissues and cerebral hemispheres were used immediately without freezing following similar steps as described in the original LysoIP protocol[12].

Protein analyses. Liver biopsy or half cerebral hemisphere samples (4 mm) were gently homogenized in 1 ml ice-cold PBS supplemented with protease and phosphatase inhibitor cocktails (Roche) using a 2 ml Dounce homogenizer (Douncer). Both liver and brain samples required 25 strokes for optimal homogenization. Homogenates were collected and 25 µl was saved on ice as the input fraction, while the remaining was centrifuged at 1,000*g* for 2 min at 4 °C. The supernatant was then incubated with anti-HA beads that were prewashed three times with the same buffer, incubated for 20 min on a gentle rotator shaker at 4 °C and then washed three times in the same cold buffer. Washes were performed by pipetting the beads three times in each wash, then discarding the supernatant by placing the tube on a magnet to trap the beads. To increase the purity of the lysosomal preparation, we changed the tube during the last wash and always used protein low binding tubes. The input and lysosomal fractions were resuspended in ice-cold Triton lysis buffer (50 mM HEPES pH 7.4, 40 mM NaCl, 1% Triton X-100, 10 mM β-glycerol phosphate, 10 mM pyrophosphate, 30 mM sodium fluoride, 1.5 mM Sodium orthovanadate, 2 mM EDTA and protease and phosphatase inhibitor cocktails). The samples were incubated for 10 min at 4 °C and centrifuged at maximum speed in cold. The supernatant was collected and saved for analysis. The lysosomal fraction was not centrifuged but rather incubated on the magnet and transferred to new tubes twice. For immunoblots, an equivalent of 0.08% and 15% of the whole tissue input and captured lysosomes were run, respectively.

Polar metabolite and lipid analyses. The procedure was similar to the steps above, but processing was performed in ice-cold KPBS (136 mM KCl, 10 mM KH_2PO_4, pH 7.25 in Optima LC–MS water) with no protease or phosphatase inhibitors. The homogenates were incubated with prewashed anti-HA beads for 5 min for polar metabolites and 20 min for lipid experiments. We noticed that incubation for 3–15 min still returns similar results for polar metabolites, but incubation for 5 min was used for convenience. For polar metabolites, the input fractions were resuspended in 225 µl, while the lysosomal fractions were resuspended in 50 µl of ice-cold 80% methanol in LC–MS water containing

500 nM isotope-labelled amino acids used as internal standards (Cambridge Isotope Laboratories). The samples were stored at −80 °C until analysed. On the day of analysis, the samples were vortexed for 10 min at 4 °C and centrifuged at 17,000g. For lipid extraction, 800 μl of chloroform:methanol at ratio of 2:1 (v/v) containing 750 ng ml^{-1} of SPLASH LipidoMIX internal standard mix (Avanti) was added to the sample, followed by the addition of 160 μl of 0.9% (w/v) NaCl and vortexing for 10 min at 4 °C. The mixture was centrifuged at 3,000g for 15 min at 4 °C. The lower phase containing lipids was collected and dried using a SpeedVac. At this stage, the samples were stored at −80 °C. On the day of analysis, dried lipid extracts were reconstituted in 50 μl of acetonitrile (ACN):isopropyl alcohol:H_2O at a ratio of 13:6:1 (v/v/v) and vortexed for 10 min at 4 °C. The samples were centrifuged for 10 min at 4 °C and transferred to glass insert vials for LC–MS.

Proteomics

Eluates from LysoIPs and aliquots of the matched starting lysates (corresponding to approximately 50 μg of protein extract) were processed as described previously[43]. In brief, proteins were solubilized by addition of SDS to a final concentration of 2% (w/v), followed by sonication using the Bioruptor Plus (Diagenode) and heating for 10 min at 95 °C. After reduction and alkylation, proteins were precipitated by cold-acetone precipitation. The resulting pellets were resuspended in digestion buffer (3 M urea in 100 mM HEPES pH 8.0) and digested by consecutive addition of LysC (Wako, 3 h at 37 °C) and trypsin (Promega, 16 h at 37 °C). The obtained digested peptides were acidified and desalted using a Waters Oasis HLB μElution Plate 30 μm (Waters) according to the manufacturer's instructions. The desalted peptides were dissolved in 5% (v/v) ACN, 0.1% (v/v) formic acid to a peptide concentration of approximatively 1 μg μl^{-1} and spiked with iRT peptides (Biognosys) before analysis using LC–MS/MS. Approximatively 1 μg of reconstituted peptides were analysed by data-independent acquisition (DIA) using the UltiMate 3000 UPLC system (Thermo Fisher Scientific) fitted with a trapping (Waters nanoEase M/Z Symmetry C18, 5 μm, 180 μm × 20 mm) and an analytical column (Waters nanoEase M/Z Peptide C18, 1.7 μm, 75 μm × 250 mm). The outlet of the analytical column was coupled directly to a QExactive HF (Thermo Fisher Scientific) system using the Proxeon nanospray source. Solvent A was water, 0.1% (v/v) formic acid and solvent B was 80% (v/v) ACN, 0.08% (v/v) formic acid. Peptides were eluted using a non-linear gradient from 1% to 62.5% B in 131 min. The total runtime was 150 min, including clean-up and column re-equilibration. The S-lens RF value was set to 60. MS acquisition parameters were set as follows: full scan MS spectra with a mass range of 350–1,650 m/z were acquired in profile mode in the Orbitrap with resolution of 120,000 FWHM. The filling time was set at a maximum of 60 ms with an AGC target of 3×10^6 ions. DIA scans were acquired with 40 mass window segments of differing widths across the MS1 mass range. The default charge state was set to 3^+. Higher energy collisional dissociation (HCD) fragmentation (stepped normalized collision energy; 25.5, 27, 30%) was applied and MS/MS spectra were acquired with a resolution of 30,000 FWHM with a fixed first mass of 200 m/z after accumulation of 3×10^6 ions or after a filling time of 35 ms (whichever occurred first). Data were acquired in profile mode. For data acquisition and processing, we used Tune v.2.9 and XCalibur v.4.1.

Acquired data were processed using Spectronaut Professional v.13.10 (Biognosys). Raw files were searched by directDIA search with Pulsar (Biognosys) against the mouse UniProt database (*Mus musculus*, entry only, release 2016_01) with a list of common contaminants appended, using the default settings. For quantification, the default BGS factory settings were used, except for the following settings: proteotypicity filter = only protein group specific; major group quantity = median peptide quantity; major group top N = OFF; minor group quantity = median precursor quantity; minor group top N = OFF; data filtering = Q value percentile with fraction = 0.2 and imputing strategy = global imputing; normalization strategy = local normalization; row selection = automatic. The candidates and protein report tables were exported from Spectronaut and were used for volcano plot generation and PCA, respectively, using R v.3.4.1 and RStudio server v.1.1.463. Protein groups were considered to be significantly enriched in LysoIP samples if they displayed a Q value of <0.01 and an average $\log_2$ ratio of >1.5. Known lysosomal proteins were defined based on Gene Ontology cellular component or UniProt subcellular localization annotation. The MS proteomics data have been deposited to the ProteomeXchange Consortium via the PRIDE[44] partner repository with the dataset identifier PXD018624.

Polar metabolomics

The initial metabolite profiling was conducted on the QExactive benchtop orbitrap mass spectrometer equipped with an Ion Max source and a heated electrospray ionization (HESI) II probe, which was coupled to a Dionex UltiMate 3000 HPLC system (Thermo Fisher Scientific). To prevent any systematic bias from possible column carryover, sample order was randomized in LC–MS runs. External mass calibration was performed using the standard calibration mixture every 7 days. Typically, 5 μl was injected onto a SeQuant ZIC-pHILIC 150 mm × 2.1 mm analytical column equipped with a 2.1 mm × 20 mm guard column (both 5 mm particle size; EMD Millipore). Buffer A was 20 mM ammonium carbonate, 0.1% ammonium hydroxide; buffer B was ACN. The column oven and autosampler tray were held at 25 °C and 4 °C, respectively. The chromatographic gradient was run at a flow rate of 0.150 ml min^{-1} as follows: 0–20 min, linear gradient from 80–20% B; 20–20.5 min, linear gradient from 20–80% B; 20.5–28 min, hold at 80% B. The mass spectrometer was operated in full-scan, polarity-switching mode, with the spray voltage set to 3.0 kV, the heated capillary was held at 275 °C, and the HESI probe held at 350 °C. The sheath gas flow was set to 40 units, the auxiliary gas flow was set to 15 units, and the sweep gas flow was set to 1 unit. MS data acquisition was performed in a range of m/z = 70–1,000, with the resolution set at 70,000, the AGC target at 1×10^6 and the maximum injection time at 20 ms. To improve sensitivity of detection of LysoTracker DND-99 and deuterated GPG (D5–GPG) in lipid-uptake experiments, additional targeted selected ion monitoring scans were included in positive mode, with m/z centred on 400.21150 and 252.08920, respectively. For these scans, the resolution was set to 70,000, the AGC targeted was 1×10^5, the max IT was 250 ms and the isolation window around each target mass was set to 2.0 m/z.

Relative quantification of polar metabolites was performed by extracting ion chromatograms with XCalibur QuanBrowser v.4.1 (Thermo Fisher Scientific) using a 5 ppm mass tolerance and referencing an in-house library of chemical standards. Metabolites and runs were subject to predefined quality-control parameters: CV (s.d./mean peak area across multiple injections of a representative (pooled) biological sample) below 0.25; and R^2 (linear correlation across a three-point dilution series of the representative (pooled) biological sample) greater than 0.90.

For a metabolite to be lysosomal in the targeted analysis performed in Fig. 1, its abundance in the immunoprecipitates from LysoTag$^+$ mice had to be twice higher than that in a control immunoprecipitate in at least 3 out of the 4 samples. Furthermore, all of the fold change calculations in the abundance of lysosomal metabolites, where targeted analyses are presented, were determined after subtracting the abundance of the metabolites in the control immunoprecipitates. For polar targeted metabolites from cultured cells, we also used LysoTracker detected with LC–MS to account for any gross changes in lysosomal mass or capturing efficiency by normalizing metabolite intensities to the total captured LysoTracker divided by the total cellular LysoTracker signal (capturing efficiency). Whole-cell abundances were normalized to total cell volume.

Note that some metabolite profiling data were acquired on an ID-X tribrid mass spectrometer (Thermo Fisher Scientific) at Stanford University. The same HILIC method and column parts were used for a Vanquish Horizon UPLC system. The different mass spectrometer parameters were as follows: Orbitrap resolution, 120,000; maximum injection time, 80 ms. Note that, due to the change of instrumentation, there is an approximately 1 min retention time shift for those polar metabolite data collected at Stanford University, hence the difference in the retention time of GPE and GPG standards from those reported in the experimental samples in Supplementary Table 4. Experimental samples were re-run and metabolite identities were validated using the new instrumentation alongside our standards.

Untargeted polar metabolomics

Data were acquired as described above, with additional data-dependent MS/MS (ddMS/MS) collected on pooled samples to help with unknown metabolite identification. For ddMS/MS, the top 10 ions in each full scan were isolated with a 1.0 Da window, fragmented with a step-wise collision energy of 15, 30 and 45 units, and analysed at a resolution of 17,500 with an AGC target of 2×10^5 and a maximum injection time of 100 ms. The underfill ratio was set to 0. The selection of the top 10 ions was set to isotopic exclusion, a dynamic exclusion window of 5.0 s and an exclusion list of background ions based on a solvent blank. Data were analysed using Compound Discoverer v.3.1 (Thermo Fisher Scientific) and by including a custom mass list. Normalization was performed using the default total ion count (TIC) parameter. Identification and annotation of metabolite species was confirmed by matching retention times and MS/MS data to authentic standards. In cases in which there was no authentic standard available, MS/MS data were matched first to published experimental MS/MS data (that is, through the Thermo Fisher Scientific mzCloud spectral library), or to predicted MS/MS data produced using CFM-ID[33]. Details of each metabolite annotated/identified in this study are provided in Fig. 3 and Supplementary Table 4.

Compound discoverer workflow. Features were extracted in the m/z = 70–1,000 mass range, from 0.1–16 min of the pHILIC chromatography runs. The signal to noise threshold for feature detection was set to 1.5. Compounds were detected using the following parameter settings: mass tolerance was set to 10 ppm, retention time tolerance was set to 0.4 min, minimum peak intensity was set to 100,000 AU and the signal to noise threshold for compound detection was set to 3. Isotopes and adducts were grouped using the default ions and base ion lists. Missing peaks were gap-filled and mock extraction blank samples were used to determine the background compounds for subsequent subtraction. Areas were normalized to constant median excluding the blank samples.

To predict elemental compositions of the compounds, the relative intensity tolerance was set to 30% for isotope pattern matching. For the mzCloud database search, both the precursor and fragment mass tolerance were set to 10 ppm. The databases used for matching compounds in ChemSpider for structural search were BioCyc, the Human Metabolome Database and KEGG, and the mass tolerance in ChemSpider Search was set to 5 ppm. The mass tolerance for matching compounds in Metabolika pathways was set to 5 ppm. Compounds were also matched to our in-house mass list, with mass tolerance set to 5 ppm and retention time tolerance set to 1 min. mzLogic was applied to score structure candidates based on MS/MS, with fragment mass tolerance set to 10 ppm and match factor threshold set to 30. Compounds were assigned by comparing annotations using the following nodes in order of priority: (1) mass list search; (2) mzCloud search; (3) predict compositions; (4) ChemSpider search; (5) Metabolika search.

The differential analysis was performed on $\log_{10}$-transformed values using defined sample ratios (*Cln3*$^{+/-}$ versus *Cln3*$^{-/-}$, both in LysoIP fraction and whole cell fraction), and *P* values were calculated. A results table of 1,073 detected compounds was generated from this workflow (Supplementary Table 4) after (1) background compound subtraction and (2) applying a filter of maximum area intensity greater than 500,000 AU to filter out low signal peaks.

The 1,073 detected compounds include possible artifacts as well as redundant or mis-annotated compounds. To focus on unique compounds with good peak quality and sufficient information for validation studies, we performed additional filtering, which included the following: (1) at least 1 compound match in the ChemSpider search; (2) at least 1 compound match in the mzCloud search. This resulted in 237 compounds. We then manually removed unreliable compounds: those with poorly aligned chromatograms and/or those that were obviously redundant. The final filtered compound list contained 189 compounds with assigned chemical formula and fragmentation results to help with compound identification (Supplementary Table 4). Among the 189 compounds after filtering, 3.2% had at least 1 $\log_2$ fold difference between the *Cln3*$^{+/-}$ versus *Cln3*$^{-/-}$ LysoIP samples, with adjusted *P* values less than 0.05. The top compounds are structurally related and can be grouped as a class of compounds termed GPDs. PCA in Fig. 2d was performed using PC1 and PC3, because PC3 includes the metabolites that separate the lysosomes of different genotypes. Whole tissues were not separated regardless of the PC used. *P* values were calculated by running the Tukey HSD test (post hoc) after ANOVA (for two group comparison, the ANOVA was simplified to unpaired *t*-tests). The *P* values were corrected using the Benjamini–Hochberg method for the false-discovery rate. These analyses were performed using the differential analysis node in compound discoverer v.3.1.

For targeted analyses, Thermo Fisher Scientific TraceFinder software v.5.0 was used to help in processing targeted metabolites screening and obtaining quantitative data from LC–MS.

In the final result table with 189 detected compounds, compound identification was performed based on accurate mass, retention time, and MS/MS spectra comparisons. We assigned to each compound one of four different levels of identification/annotation according to guidelines issued by the MSI[45]. Level 1 defines compounds identified with both experimental retention time and MS/MS fragmentation patterned matched to authentic chemical standards that were run under the same chromatographic conditions. Level 2 defined putatively annotated compounds, with only one of the orthogonal properties (such as retention time) matched with an authentic chemical standard, or an MS2 match to only the online in silico spectra library or the Thermo Fisher Scientific mzCloud database. Level 3 defined putatively characterized compound classes. Finally, level 4 defined unknown compounds.

Lipidomics

Lipids were separated on an Ascentis Express C18 2.1 mm × 150 mm 2.7 µm column (Sigma-Aldrich) connected to a Vanquish Horizon UPLC system and an ID-X tribrid mass spectrometer (Thermo Fisher Scientific) equipped with a HESI probe. External mass calibration was performed using the standard calibration mixture every 7 days. Between 2 µl and 5 µl of sample were injected onto the column, with separate injections for positive and negative ionization modes. Mobile phase A in the chromatographic method consisted of 60:40 water:ACN with 10 mM ammonium formate and 0.1% formic acid, and mobile phase B consisted of 90:10 isopropanol:ACN, with 10 mM ammonium formate and 0.1% formic acid. The chromatographic gradient was adapted from ref. [46]. In brief, the elution was performed with a gradient of 40 min; during 0–1.5 min isocratic elution with 32% B; from 1.5 to 4 min, increase to 45% B; from 4 to 5 min, increase to 52% B; from 5 to 8 min, increase to 58% B; from 8 to 11 min, increase to 66% B; from 11 to 14 min, increase to 70% B; from 14 to 18 min, increase to 75% B; from 18 to 21 min, increase to 97% B; during 21 to 35 min, 97% B is maintained; from 35–35.1 min, solvent B was decreased to 32% and then maintained for another 4.9 min for column re-equilibration. The flow rate was set to 0.260 ml min^{-1}. The column oven and autosampler were held at 55 °C and 15 °C, respectively. The mass spectrometer parameters were as follows: the spray voltage was set to 3.25 kV in positive mode and 3.0 kV in negative mode, and the heated capillary and the HESI were held at 300 °C and 375 °C, respectively. The S-lens RF level was set to 45, and the sheath and auxillary gas were set to 40 and 10 units, respectively. These conditions were held constant for both positive- and negative-ionization mode acquisitions.

The mass spectrometer was operated in full-scan ddMS/MS mode with an orbitrap resolution of 120,000 (MS1) and 30,000 (MS/MS). Internal calibration using Easy IC was enabled. Quadrupole isolation was enabled, the AGC target was 1×10^5, the maximum injection time was 50 ms and the scan range was $m/z = 200$–2,000. For ddMS/MS, the cycle time was 1.5 s, the isolation window was 1 and an intensity threshold of 1×10^3 was used. HCD fragmentation was achieved using a step-wise collision energy of 15, 25 and 35 units, and detected in the orbitrap with an AGC target of 5×10^4 and a maximum injection time of 54 ms. Isotopic exclusion was on, a dynamic exclusion window of 2.5 s was used and a top 100 exclusion list was generated using a solvent bank.

High-throughput annotation and relative quantification of lipids was performed using LipidSearch v.4.2.21 (Thermo Fisher Scientific/ Mitsui Knowledge Industries)[47,48] using the HCD and 'Labelled' databases (the latter was searched to detect the Avanti Splash LipidoMix that was used as an internal standard). Precursor ion tolerance was set to 5 ppm and product ion tolerance was set to 8 ppm. LipidSearch nomenclature uses underscores to separate the fatty acyl chains to indicate the lack of *sn* positional information (for example, PC(16:0_18:1)) and not (PC(16:0/18:1)). Raw peak areas for all annotated lipids were exported to Microsoft Excel and filtered according to the following predetermined quality control criteria: Rej (reject parameter calculated by LipidSearch) equal to 0; PQ (peak quality parameter calculated by LipidSearch software) greater than 0.75; CV (s.d./mean peak area across triplicate injections of a represented (pooled) biological sample) below 0.4. Raw peak areas of the filtered lipids were normalized to well-detected deuterated standards found in the Avanti Splash LipidoMix as a control for lipid extraction and efficiency, as well as sample loading. Individual lipid species shown in figures were validated by manually checking the peak alignment and matching the MS/MS spectra to the characteristic fragmentation patterns found in the LIPID-MAPS database (www.lipidmaps.org). Notably, BMP lipid class is not included in the Lipidsearch target database—the PG lipid species from Lipidsearch analysis might be misannotated because BMP and PG are structural isomers. To distinguish between PG and BMP lipid species, annotation for those two lipid classes was performed manually based on the retention time and positive mode MS2 fragmentation. The lipid standards (18:1 (Δ9-Cis) PG and 18:1 BMP (S,R)) were purchased from Avanti Polar Lipids and found to be separable under our lipid chromatographic conditions. The BMP standard elutes before its PG isomer, either when the pure standards were prepared in isopropanol:methanol (1:1) solvent or when spiked into a bovine liver lipid extract. Although BMP and PG have better sensitivity in negative-ionization mode, their positive ammonium adducts were useful precursors for generating distinct neutral loss fragments critical for annotation. Consistent with a previous report[49], the PG standard generated loss of the polar head group leading to a diglyceride-like fragment. By contrast, the BMP standard generated loss of the monoacylglycerolphosphate moiety leading to monoglyceride-like fragments. When MS2 fragmentation was not acquired, the species was annotated as BMP/PG.

Protein extractions from mouse tissues

Immediately after euthanasia, tissues were collected and snap-frozen in liquid nitrogen. Before protein extraction, tissues were powdered in liquid nitrogen using a mortar and a pestle. RIPA lysis buffer (50 mM Tris-HCl, pH 8.0, 150 mM NaCl, 0.1% Triton X-100, 0.5% sodium deoxycholate, 0.1% SDS) supplemented with protease and phosphatase inhibitors was added to 25 mg of tissue powder and the samples were incubated on shaker for 1 h at 4 °C. Lysates were centrifuged for 30 min at maximum speed and the protein concentrations were determined using the Pierce BCA Protein Assay kit (Thermo Fisher Scientific).

Immunoblotting

Lysates were resolved by SDS–PAGE (Thermo Fisher Scientific) at 120 V. Resolved proteins were transferred for 2 h at 40 V to PVDF membranes. Membranes were blocked with 5% non-fat dry milk in TBST (Tris-buffered saline with Tween-20) for 1 h, then incubated overnight with primary antibodies in 5% BSA in TBST at 4 °C. A list of all primary antibody dilutions is provided in Supplementary Table 6. After incubation, membranes were washed three times with TBST for 5 min per wash and then incubated with the appropriate secondary antibodies diluted 1:3,000 in 5% BSA for 1 h at room temperature. Membranes were then washed three times with TBST and visualized using ECL2 western blotting substrate (Thermo Fisher Scientific).

Electron microscopy

For electron microscopy studies, mice were perfused with 2% paraformaldehyde and 2.5 % glutaraldehyde. Brains and livers were then isolated and postfixed in the same fixing solution. Coronal 200 μm sections were cut using the Leica VT-1000 (Leica) vibratome. Sections were post-fixed with 2% osmium, rinsed, dehydrated and embedded in Durcupan resin (Sigma-Aldrich). Ultra-thin sections (0.07–0.09 μm) were cut with a diamond knife with an ultramicrotome (Leica EM UC6, Leica), stained with lead citrate (Reynolds solution) and examined with a transmission electron microscope (FEI Tecnai G2 Spirit BioTwin, Thermo Fisher Scientific) using a digital camera (Morada, Olympus Soft Image Solutions).

Cathepsin B activity assay

Cathepsin B activity was assayed fluorometrically using 5 μM of the fluorogenic CTSB substrate, Z-RR-AMC (Sigma) in 100 μl of the cathepsin B activation buffer (KH_2PO_4 (87.7 mM), Na_2HPO_4 (12.3 mM), EDTA (4 mM), DTT (2.6 mM) and cysteine (8 mm) in pH 6.0). The plate was incubated in the Tecan Spark multimode microplate reader (Tecan Trading) at 37 °C and read at an excitation of 380 nm and emission of 460 nm every 5 min. The relative fluorescence unit (RFU) values of all of the samples were corrected by subtracting the value of a blank sample.

β-Hexosaminidase activity assay

β-Hexosaminidase activity was assayed using a *N*-acetyl-β-D-glucosaminide (sugar-substrate) conjugated to the fluorophore 4-methylumbelliferone (4-MU). 4-MU–*N*-acetyl-β-D-glucosaminide (0.6 mM) in 50 mM sodium acetate buffer, pH 5.4, was used with 5 μg of brain or liver homogenates and incubated at 37 °C for several time points. The reaction was stopped by adding 0.1 M glycine buffer, pH 10. The fluorescence of released 4-MU was measured in a Tecan Spark multimode microplate reader (Tecan Trading) at an emission wavelength of 448 nm and excitation at 362 nm. The RFU values of all of the samples were corrected by subtracting the value of a blank sample.

Immunofluorescence assays in cells and tissues

Cells. A total of 50,000 cells were plated on fibronectin-coated glass cover slips, fixed with 4% paraformaldehyde and blocked with 5% donkey serum in PBS for 1 h at room temperature. This was followed by overnight incubation at 4 °C with primary antibodies. The primary antibodies used in this study were mouse anti-LAMP2 (1:500) and rabbit anti-Flag (1:500). Secondary antibodies (Alexa-Fluor) were then applied in 1% BSA, 0.3% Triton X-100 for 2 h. Images were acquired in *z*-series on a spinning-disk confocal system and maximum-intensity projections were processed using ImageJ v.1.52.

Tissues. Immediately after euthanasia, brains and livers were removed and fixed by immersion in 4% paraformaldehyde overnight, and embedded in paraffin. For experiments including *Cln3*-knockout mice, after fixation, tissues were transferred to 30% sucrose overnight, embedded in OCT (TissueTek) and frozen. For immunofluorescence staining of paraffin-embedded tissues, 5 μm sections were deparaffinized. Heat-induced antigen retrieval was performed in Borg Decloaker (Biocare Medical) buffer using a pressure cooker (Dako). For cryo-sections, antigen retrieval was omitted. Sections were blocked in 5% donkey serum, 0.3% Triton X-100 for 2 h. Primary antibodies (rat anti-LAMP1, mouse anti-HA and rabbit anti-NeuN) were applied in 1% BSA, 0.3% Triton X-100 for 48 h. Secondary antibodies were then applied in 1% BSA, 0.3% Triton X-100 for 2 h. The sections were covered by cover slips with Vectashield (Vector Laboratories). Images were acquired in *z*-series on a spinning-disk confocal system.

Images were acquired using the Airyscan2 LSM980 microscope or Zeiss AxioVert200M microscope using ZEN v.3.2 imaging software or MetaMorph 7, respectively, and maximum-intensity projections were processed using ImageJ v.1.52.

Cell lines and plasmids

The pMXs-IRES-Bsd vector was obtained from Cell Biolabs. HEK293T cells were purchased from ATCC and authenticated by STR profiling. The sequence of the human *CLN3* cDNA was mutated to remove the proto-spacer adjacent motif, as well as within the seed region of the sgRNA used in this study. Other mutations were also introduced to the sequence targeting another sgRNA site that was not included in the current study.

Cell culture

HEK293T cells and their derivatives were cultured in DMEM base medium with 10% inactivated fetal calf serum (Thermo Fisher Scientific) supplemented with 2 mM glutamine, penicillin and streptomycin (Thermo Fisher Scientific). All of the cell lines were maintained at 37 °C and 5% CO_2. For all LysoIP experiments using HEK293T cell lines, cells were cultured in RPMI base medium supplemented with 4 nM of LysoTracker Red DND-99 for 1 h before processing as described in ref. [12].

Virus production and transduction

Lentiviruses were produced by transfecting HEK293T cells with lentiviral plasmids of the lysosomal tags in combination with the packaging plasmids VSV-G env elope and ΔVPR. For retrovirus production, pMXs Flag-CLN3 plasmid and retroviral packaging plasmids Gag-Pol and VSV-G were co-transfected using XtremeGene9 transfection reagent. The culture medium was changed to DMEM supplemented with 30% inactivated fetal calf serum 16 h after transfection. The virus-containing supernatant was collected 48 h after transfection and centrifuged for 5 min at 400*g* to remove cells and then frozen at −80 °C. To establish cell lines with stable expression, 500,000 WT or *CLN3*-knock-out HEK293T cells were plated in six-well plates in DMEM with 10% inactivated fetal calf serum and 8 μg ml^{-1} polybrene and 100–250 μl of virus-containing medium. Spin-infection was performed at 2,200 rpm for 45 min at 37 °C. Cells with the virus were then incubated for 16 h before adding fresh culture medium containing blasticidin (InvivoGen) for selection for at least 72 h . Stable cell lines were tested for proper localization of the lysosomal tag or Flag–CLN3 using immunofluorescence.

Generation of knock-out cell lines using CRISPR–Cas9 technology

Human CLN3 was depleted using the pLentiCRISPRv1 system. The following sense (S) and antisense (AS) oligo-nucleotides were cloned into pLentiCRISPRv1: (Upper case bases are for target gene and lower case bases are for cloning) sgCLN3 (S): caccgGTCACGCTCAGCAATACCGC; sgCLN3 (AS): aaacGCGGTATTGCTGAGCGTGACc; sgAAVS (S): caccgTCCCCTCC ACCCCACAGTG; and sgAAVS (AS): aaacCACTGTGGGGTGGAGGGGAc.

To generate knock-out cells, HEK293T cells were transduced as described above and after puromycin selection, cells were single-cell sorted into 96-well plates containing 200 μl of DMEM supplemented with 30% inactivated fetal calf serum. Cell clones with the desired knock-outs were identified by Illumina amplicon deep sequencing at Massachusetts General Hospital. Control cells were generated by targeting the non-coding *AAVS1* locus as described previously[50].

Generation and purification of GPG and GPE

To obtain GPG and GPE, phosphatidylglycerol (16:0 PG) and phosphatidylethanolamine (16:0 PE) (Avanti) were saponified as follows: 25 mg of each was dissolved in a 3 ml 2:1 chloroform:methanol solution in a clean 20 ml scintillation vial provided with a magnetic stirring bar. NaOH (1 ml of 2 M) was added to the mixture and the reaction was stirred at room temperature for 2 h. HCl (1 ml of 2 M) was used to neutralize the reaction and GPDs were extracted with 2 ml water twice into separate vessels. Aqueous phases were combined, washed with 2 ml chloroform, frozen in liquid nitrogen and lyophilized. To separate the crude product from inorganic salts, the lyophilates were dissolved in a minimum amount of methanol, and the desired products were separated from the precipitants by paper filtration.

Metabolite extraction from yeast

Saccharomyces cerevisiae (WT strain BY6741 and *Btn1Δ* strain P35H10) were obtained by the Jarosz laboratory at Stanford University. Yeast strains were grown in yeast nitrogen base (with amino acids and 2% glucose) and collected during the early growth phase (OD_{600} of 0.5–1.0). Cells were centrifuged for 3 min at 2,700*g* at the lowest temperature and the cell pellets were washed twice with cold PBS. Cells were then resuspended in 1 ml of prechilled quenching solution (2:2:1 mixture of ACN:MeOH:H_2O) and transferred to 2 ml vials half filled with 0.5 mm glass beads. The cells were homogenized by 10 pulses of 10–20 s each at 3,800–5,000 rpm with 2 min rest on ice between each pulse. The homogenates were removed and centrifuged at 2,700*g* in the cold for 2 min. The supernatants were removed to prechilled tubes and centrifuged for 30 min at 17,000*g* in the cold. The supernatants were transferred to new tubes and stored at −80 °C for subsequent analysis using LC–MS. This protocol was adapted from ref. [51]

Glycerophosphodiester phosphodiesterase activity assay

HEK293T cells stably expressing 3×Flag–GDE or 3×Flag–CLN3 were collected and homogenized on ice. The resulting supernatant was immunoprecipitated with anti-Flag-conjugated agarose beads for 4 h at 4 °C. Tagged proteins were competitively eluted using Flag peptide. The activity assay was carried out in 50 μl buffer containing 100 mM Tris (pH 7.5), 10 mM $MgCl_2$, 2 μg fatty-acid free BSA, 50 μM labelled D9–GPC, and 2 μg of 3×Flag–GDE or 3×Flag–CLN3 protein[52]. The incubation was performed at 37 °C and terminated by the addition of 80% cold methanol. GDE activity was assessed by the change in the level of D9–GPC and the generation of D9–choline using MS analysis.

Measurement of lysosomal pH

Lysosomal pH was measured using the LysoSensor Yellow/Blue DND-160 indicator dye. In brief, the cells were incubated with 5 μM LysoSensor Yellow/Blue DND-160 in complete cell growth medium for 10 min at 37 °C. The cells were then washed three times with PBS. The cells were subsequently analysed using microplate reader and fluorescence emissions were collected at 440 nm and 540 nm for excitations at 340 nm and 380 nm, respectively. LysoSensor Yellow/Blue is ratiometric, so the ratio of light excited at 340 nm to 380 nm is proportional with lysosomal pH that is measured using a calibration curve. To create a standard curve, the cells with LysoSensor were equalized by incubating them in a series of MES calibration buffer (20 mM MES (2-(*N*-morpholino) ethanesulfonic acid), 110 mM KCl, 20 mM NaCl, 10 μM monesin, 20 μM nigericin) at pH values ranging from 3.5 to 7.5 for 10 min at 37 °C.

In vitro lipid-uptake assay

Fatty-acid-free BSA (1.25%; FAF-BSA) was dissolved in DMEM medium at 37 °C and subsequently filtered. Then, 0.1 mM of 16:0–18:1 D5–PG lipid (Avanti) was incubated with FAF-BSA at room temperature for 30 min on a rotating shaker. The molar ratio of D5–PG to FAF-BSA was 1:1. This mixture of D5–PG conjugated to FAF-BSA was considered as 10× stock and experiments were performed at a 1× working concentration. Before supplementation with D5-PG:FAF-BSA, cells were serum-starved for 2 h to enhance endocytosis. Cells were then incubated with the mixture for 4 h. Treated and untreated cells were subjected to LysoIP followed by LC–MS analysis as described above.

D9-dipalmitoylphosphatidylcholine (16:0–16:0) and D9-choline labelling in primary neurons

Cln3$^{+/-}$ and *Cln3*$^{-/-}$ postnatal (day 1) mouse pups were euthanized by decapitation, and the cortices were collected by microdissection. Fresh cortices were then washed by dissection medium (Thermo Fisher Scientific, 14170161) before tissue dissociation using scissors, 0.25% trypsin and pipette trituration. Cell strainers (70 μm) were used to remove tissue chunks from digested cortices. Dissociated neurons were plated with minimum essential medium (Thermo Fisher Scientific, 21010046) containing 10% inactivated fetal calf serum (Thermo Fisher Scientific, 10438026) supplemented with 2 mM glutamine, penicillin and streptomycin (Thermo Fisher Scientific, SV30010) in 24-well plates with coverslips coated with poly-L-lysine (Newcomer Supply, 1339A). A complete medium change to neurobasal medium (Thermo Fisher Scientific, 21103049) supplemented with B-27 (Invitrogen, 17504044) and 2 mM GlutaMax (Thermo Fisher Scientific, 35050061) was performed 24 h after initial seeding. Primary neuronal cells were maintained at 37 °C and 5% CO_2. Neurons (day 7) were first starved in neurobasal medium without B-27 for 30 min, and were then labelled with 37.6 μM d9-dipalmitoylphosphatidylcholine (16:0–16:0) conjugated with 0.25% fatty acid-free bovine serum albumin (BSA) or 37.6 μM d9-choline chloride in neurobasal medium and B-27 for 24 h. Molar percent enrichment for trimethyl-d9 labelling was calculated based on the following equation:

$$\mathrm{MPE} = \frac{\sum_{i=0}^{n} i \times m_i}{n \times \sum_{i=0}^{n} m_i} \times 100\%$$

where n is the number of hydrogen atoms, mi the abundance of a mass isotopomer and i the labelling state (M+i). As a choline-containing metabolite can only be present as unlabelled (M+0) or D9-labelled (M+9), the equation can be simplified as follows:

$$\mathrm{MPE} = \frac{\text{D9-labelled species}}{\text{D9-labelled species + unlabelled species}} \times 100\%$$

Human studies

Ethics statement. The Eunice Kennedy Shriver National Institute of Child Health and Human Development, National Institutes of Health IRB approved the natural history studies. Individuals of any age, sex and race who have at least one variant in *CLN3* and clinical symptoms suggestive of CLN3 disease were enrolled following appropriate provision of consent and assent. For children, parents or legally authorized representatives provided permission and consent.

CSF sample collection. Collection of CSF samples used in this study was as previously described[35]. We performed lumbar puncture under sedation, in participants unable to cooperate with the procedure when awake, to collect CSF from individuals with CLN3 (Investigations of Juvenile Neuronal Ceroid Lipofuscinosis (CLN3); NCT033007304), creatine transporter deficiency (NCT02931682) or Smith–Lemli–Opitz syndrome (NCT00001721). Anonymized residual CSF samples from outside clinical laboratories were also collected (NCT00344331). Participants in the CLN3 natural history study have a consistent identifier (SP_._._) across publications to facilitate data comparisons. These are included in Supplementary Table 5.

Data preparation and statistics

Displays of quantitative data were prepared in Microsoft Excel v.16.61 and 16.62 and GraphPad Prism v.9.0. Statistical comparisons were performed using two-tailed unpaired t-tests in Prism or Excel, unless stated otherwise in the figure legends. All of the displayed measurements represent samples generated independently or biological replicates. Immunoblot and immunofluorescence data are representative experiments and the number of independent experiments performed is indicated in the legend. Drawings used in creating the illustrations in Fig. 1a,c and Extended Data Figs. 2a and 4a,e were created using BioRender and were used with permission.

Reporting summary

Further information on research design is available in the Nature Research Reporting Summary linked to this article.

Data availability

The MS proteomics data were deposited to the ProteomeXchange Consortium through the PRIDE partner repository with the dataset identifier PXD018624. Known lysosomal proteins in our proteomics analysis were defined based on Gene Ontology Cellular Component or UniProt subcellular localization annotation (https://www.uniprot.org/). The conditional LysoTag mouse was deposited at the Jackson Laboratory (strain 035401). Other unique biological materials in the form of plasmids or cell lines are available from the corresponding author on request. Detailed lipidomics and metabolomics data are provided in Supplementary Tables 3 and 4, respectively. Other data generated are available from the corresponding author on request. Source data are provided with this paper.

43. Wyant, G. A. et al. NUFIP1 is a ribosome receptor for starvation-induced ribophagy. *Science* **360**, 751–758 (2018).
44. Perez-Riverol, Y. et al. The PRIDE database and related tools and resources in 2019: improving support for quantification data. *Nucleic Acids Res.* **47**, D442–D450 (2019).
45. Salek, R. M., Steinbeck, C., Viant, M. R., Goodacre, R. & Dunn, W. B. The role of reporting standards for metabolite annotation and identification in metabolomic studies. *Gigascience* **2**, 13 (2013).
46. Bird, S. S., Marur, V. R., Sniatynski, M. J., Greenberg, H. K. & Kristal, B. S. Serum lipidomics profiling using LC-MS and high-energy collisional dissociation fragmentation: focus on triglyceride detection and characterization. *Anal. Chem.* **83**, 6648–6657 (2011).
47. Taguchi, R. & Ishikawa, M. Precise and global identification of phospholipid molecular species by an Orbitrap mass spectrometer and automated search engine Lipid Search. *J. Chromatogr. A* **1217**, 4229–4239 (2010).
48. Yamada, T. et al. Development of a lipid profiling system using reverse-phase liquid chromatography coupled to high-resolution mass spectrometry with rapid polarity switching and an automated lipid identification software. *J. Chromatogr. A* **1292**, 211–218 (2013).
49. Hankin, J. A., Murphy, R. C., Barkley, R. M. & Gijon, M. A. Ion mobility and tandem mass spectrometry of phosphatidylglycerol and bis(monoacylglycerol)phosphate (BMP). *Int. J. Mass Spectrom.* **378**, 255–263 (2015).
50. Wang, T., Wei, J. J., Sabatini, D. M. & Lander, E. S. Genetic screens in human cells using the CRISPR-Cas9 system. *Science* **343**, 80–84 (2014).
51. Krink-Koutsoubelis, N. et al. Engineered production of short-chain acyl-coenzyme A esters in *Saccharomyces cerevisiae*. *ACS Synth. Biol.* **7**, 1105–1115 (2018).
52. Zheng, B., Berrie, C. P., Corda, D. & Farquhar, M. G. GDE1/MIR16 is a glycerophosphoinositol phosphodiesterase regulated by stimulation of G protein-coupled receptors. *Proc. Natl Acad. Sci. USA* **100**, 1745–1750 (2003).

Acknowledgements We thank the members of the Abu-Remaileh and Sabatini laboratories for insights; J. M. Kirkpatrick and the FLI Core Proteomics; and the participants and families in this study for their invaluable contributions. This work was supported by grants from the NIH (DP2-CA271386), Stanford Alzheimer's Disease Research Center (ADRC), Charles King fellowship, BeatBatten (Netherlands) and NCL-Stiftung (Germany) foundations to M.A.-R. and from the NIH (R01 CA103866 and R01 CA129105) to D.M.S. N.N.L. and W.D. were also supported by the BeatBatten (Netherlands) and NCL-Stiftung (Germany) foundations; U.N.M. by the Stanford ChEM-H Chemistry/Biology Interface Program, O'Leary-Thiry Graduate Fellowship and NIH T32 training grant (T32GM120007); A.L.C. by the F31 predoctoral NRSA (5F31DK113665); and C.G. and R.T. were partially supported by funding from the Deutsche Forschungsgemeinschaft (DFG, German Research Foundation), Projektnummer 23928380—TRR 152 and the NCL-Stiftung (Germany); A.O. by DFG through the Research Training Group ProMoAge (GRK 2155), the Else Kröner Fresenius Stiftung (no. 2019_A79), the Fritz-Thyssen foundation (no. 10.20.1.022MN) and the Chan Zuckerberg Initiative Neurodegeneration Challenge Network (nos 2020-221617 and 2021-230967). The FLI is a member of the Leibniz Association and is financially supported by the Federal Government of Germany and the State of Thuringia. The CLN3 natural history study is funded by an NIH Clinical Center Bench to Bedside award and the NICHD intramural research programme (ZIA HD008989). M.A.-R. is a Terman Faculty Fellow at Stanford University.

Author contributions M.A.-R. and D.M.S. initiated the project and, with help from N.N.L., designed the research plan. N.N.L. and W.D. performed most of the experiments and analysed the data with help from A.L.C., V.D. and S.H.C.; S.H.C., T.K. and C.A.L. had a critical role in LC–MS analysis. W.D. also helped with analysing LC–MS data and U.N.M. generated GPD chemical standards. I.H. and A.O. designed the proteomic experiments and analysed the lysosomal proteomic data. R.T. and C.G. were consulted on CLN3 function and edited the manuscript. A.N.D.D. and F.D.P. supervised the CLN3 natural history study and provided the CSF samples. M.A.-R. and N.N.L. wrote the manuscript and D.M.S. edited it.

Competing interests M.A.-R. is a scientific advisory board member of Lycia Therapeutics. A.N.D.D. and F.D.P. have a collaborative research agreement with Amicus Therapeutics. The other authors declare no competing interests.

Additional information

Supplementary information The online version contains supplementary material available at https://doi.org/10.1038/s41586-022-05221-y.

Correspondence and requests for materials should be addressed to Monther Abu-Remaileh.

Peer review information *Nature* thanks David Pearce, Peter Vangheluwe and the other, anonymous, reviewer(s) for their contribution to the peer review of this work.

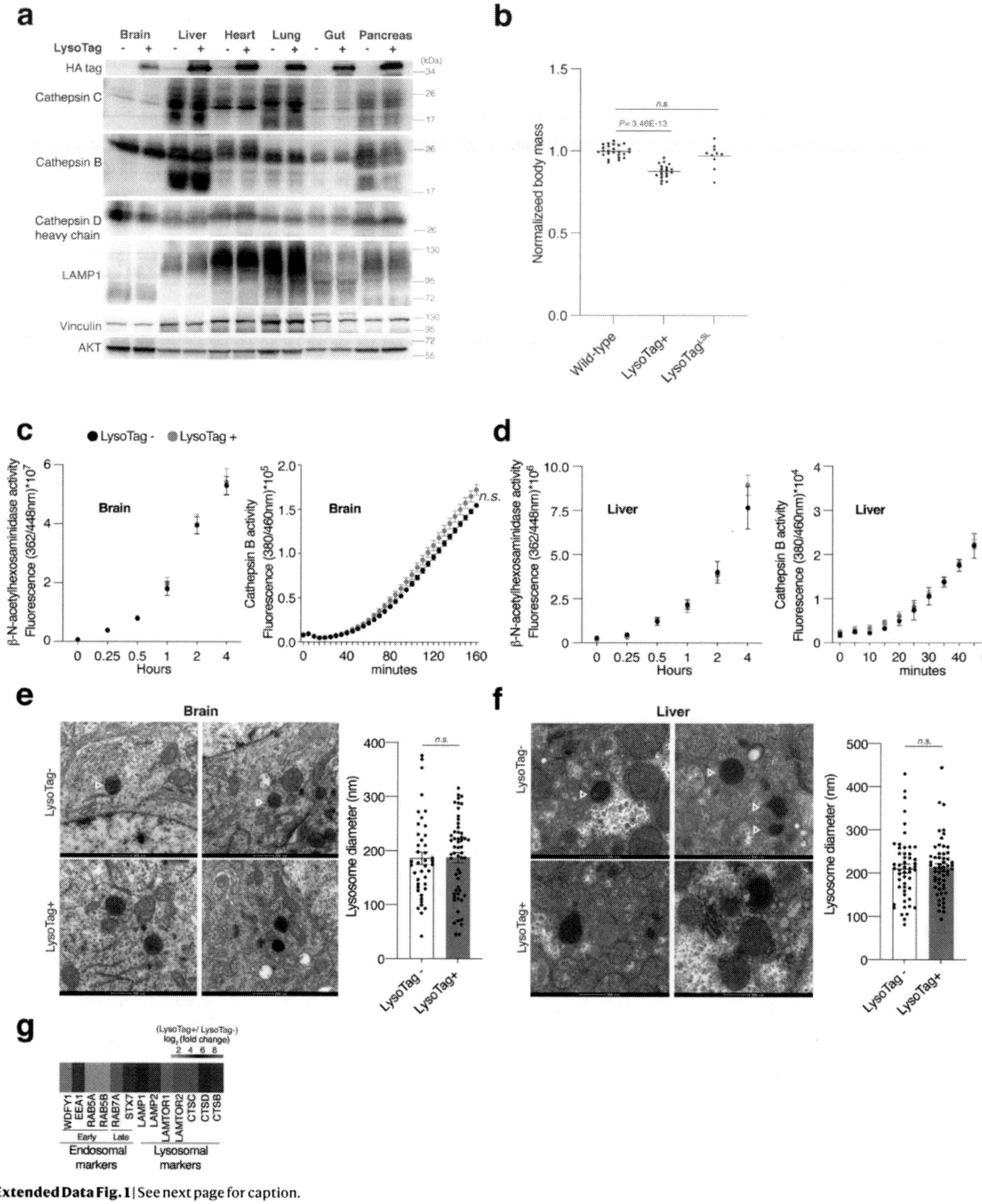

Extended Data Fig. 1 | See next page for caption.

Extended Data Fig. 1 | The effects of LysoTag expression on lysosomes in vivo. **a**, In the constitutive LysoTag mice, TMEM192-3×HA is expressed across all tissues examined as determined by immunoblotting using antibodies to the HA epitope. No difference in the expression of lysosomal markers LAMP1, Cathepsin B, C, and D is observed upon the expression of TMEM192-3×HA. Vinculin and AKT were used as loading controls. Numbers indicate molecular weights in kDa according to protein standards run on the same gel. Immunoblot is representative of two independent experiments. **b**, Normalized body weights of mice with the indicated genotypes at 5-9 weeks of age. Mice were weighed individually. The body mass of each mouse was normalized to the mean mass of sex-matched wild-type littermates before pooling by genotype for statistical analyses using a two-tailed unpaired *t*-test. ($n = 25$, 19, and 10 for Wild type, LysoTag+, and LysoTag^{LSL} mice, respectively.). **c**,**d**, Measurement of lysosomal hydrolase activity in brain and liver tissues of the constitutive LysoTag mice shows no effect of TMEM192-3×HA expression on lysosomal function. For measuring Cathepsin B activity, its fluorogenic substrate was incubated with homogenates of the corresponding tissues (see methods). For determining β-hexosaminidase activity, fluorogenic 4-methylumbelliferyl (MUF)- N-acetyl-β-Glucosaminide was used as a substrate (see methods). Fluorescence was measured at 37 °C and data were shown following the subtraction of background fluorescence observed in the absence of homogenate (mean ± s.e.m.; n=4). **e**,**f**, Representative transmission electron micrographs showing lysosomes from brain and liver tissues of LysoTag- (control) and LysoTag+ mice. Lysosomes were identified by the presence of a single membrane and granular, electron-dense appearance. White and Red arrowheads indicate lysosomes in cells from LysoTag- and LysoTag+ mice, respectively. $n = 2$ mice per group and scale bar, 500 nm. The graph showing the mean diameter of lysosomes in multiple quantified cells. The diameter of lysosome in each group was generated by measuring lysosomes using ImageJ v1.52. (For brain, $n = 42$ and $n = 55$ measured lysosomes from LysoTag- and LysoTag+ mice, respectively. For liver, $n = 51$ and $n = 56$ measured lysosomes LysoTag- and LysoTag+ mice, respectively. Data presented as mean ± s.e.m. n.s: non-significant; Two-tailed unpaired *t*-test (**e**,**f**). **g**, Heatmap presentation of the enrichment of early- and late-endosomal markers as well as those for lysosome in the LysoIP performed from mouse liver tissue. Enrichment values are derived from Supplementary Table 1. For gel source data, see Supplementary Fig. 1.

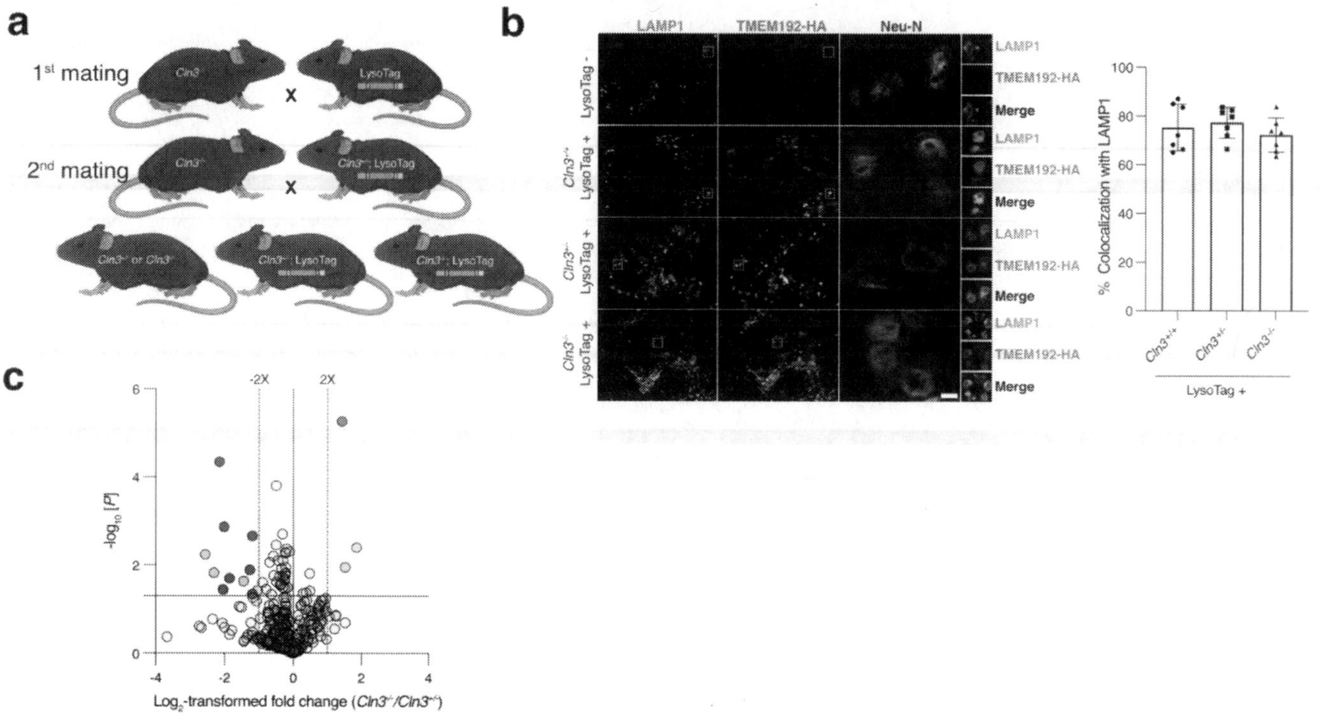

Extended Data Fig. 2 | Generation and validation of LysoTag mouse model for Batten disease studies. **a**, Schematic depicting the breeding strategy to generate LysoTag models for Batten disease studies. *Cln3*$^{-/-}$ mice were crossed with heterozygous constitutive LysoTag mice and their progeny were then crossed back to *Cln3*$^{-/-}$ animals to generate the indicated experimental mouse genotypes. Mouse drawing was created with BioRender. **b**, Expression and localization of the TMEM192-3×HA fusion protein to lysosomes in brain neurons of *Cln3*$^{+/+}$, *Cln3*$^{+/-}$ *and Cln3*$^{-/-}$ LysoTag mice. TMEM192-3×HA, lysosomes and neurons were detected in an immunofluorescence assay using antibodies to the HA epitope the lysosomal marker, LAMP1 and the neuronal marker Neu-N, respectively. Scale bar = 5 µm, and magnified insets are labelled with white boxes in the main image. Percentage of colocalization between TMEM192-3×HA and LAMP1 shown in the right panel was measured in $n = 7$ cells per genotype. Data are shown as mean ± s.e.m. Micrographs are representative of at least three independent experiments. **c**, Volcano plot comparing the lipidomic data from whole brain homogenates of *Cln3*$^{+/-}$ and *Cln3*$^{-/-}$ mice shows that only few lipids change significantly and they belong to phospholipids (yellow) and lysophosphatidylglycerol (LPG) in red. Loss of CLN3 also led to a decrease in the brain of bis(monoacylglycero)phosphate (BMP), a class of lysosomal lipids (blue). BMP/PG (cyan) annotation was used when MS/MS fragmentation was not acquired (see Supplementary Table 3 and methods for details). Horizontal line indicates a p-value of 0.05, and vertical dotted lines a fold change of 2. Each dot represents a lipid species. Data were acquired in negative ion mode and normalized to internal lipid standards for the best-matched lipid class ($n = 5$ and 4 for *Cln3*$^{+/-}$ and *Cln3*$^{-/-}$ mice, respectively). Female mice with an average age of 7 months were used. Data are in Supplementary Table 3. Two-tailed unpaired *t*-test was used.

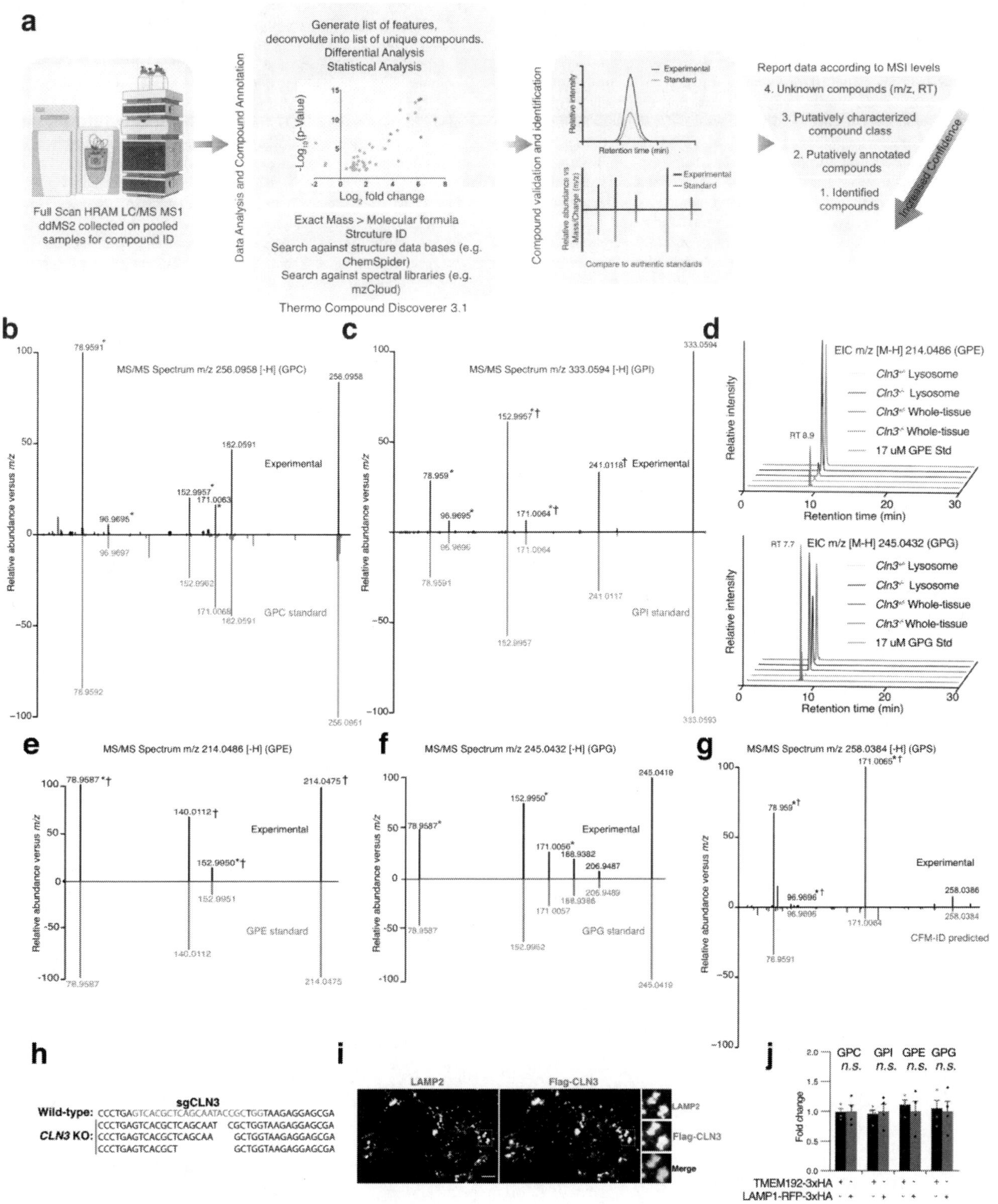

Extended Data Fig. 3 | See next page for caption.

Extended Data Fig. 3 | Validation of glycerophosphodiesters (GPDs) as the metabolites that accumulate in lysosomes upon CLN3 loss. a, Identification and annotation of metabolites using untargeted metabolomics analyses. Overview of our untargeted metabolomics experimental workflow. MS1 data were collected for every sample using full scan mode on a high-resolution accurate-mass instrument. MS/MS data were collected from pooled samples only to aid with compound identification. Data were then analysed using Thermo Compound Discoverer v3.1 to generate a list of unique features which were deconvoluted into a list of unique compounds. Differential and statistical analyses were performed. Compounds which were statistically significant in the differential analyses were validated and identified using authentic standards, when available. If no authentic standard was available, compounds were annotated by comparing MS/MS data to a database, or *in silico* fragmentation prediction (competitive fragmentation modelling-ID, CFM-ID). Data were then reported according to the Metabolomics Standards Initiative (MSI) guidelines. **b**,**c**, Mirror plots for GPC and GPI in negative ion mode, respectively. Fragments common to the glycerol-phosphate group are indicated with an asterisk*. Fragments common to those found in the MS/MS spectra reported by Kopp et al.[32] are indicated with †. **d**, EICs for GPE and GPG across a range of samples were analysed alongside in-house generated standards (Std) and showed a matching retention time (RT). **e**,**f**, Mirror plots for GPE and GPG, respectively. Fragments common to the glycerol-phosphate group are indicated with an asterisk*. Fragments common to those found in the MS/MS spectra reported by Kopp et al.[32] are indicated with †. **g**, Mirror plots showing MS/MS spectra for glycerophosphoserine from experimental samples compared to MS/MS spectra generated *in silico* using CFM-ID. Fragments common to the glycerol-phosphate group are indicated with an asterisk*. Fragments common to those found in the MS/MS spectra reported in Kopp et al.[32] are indicated with †. **h**, Amplicon sequencing was used to validate *CLN3* knock out cells generated using CRISPR-Cas9. Three mutant alleles were identified and all were found to generate frameshift deletions in the *CLN3* coding sequence as compared to the wild-type sequence. **i**, Localization of Flag-CLN3 protein to lysosomes. Flag-CLN3 and lysosomes were detected in an immunofluorescence assay using antibodies to the Flag epitope and the lysosomal marker LAMP2, respectively. Scale bar = 5 μm. Micrographs are representative images of three experiments. **j**, Changes in GPD levels in CLN3-deficient cells expressing TMEM192-3×HA tag or LAMP1-3×HA tag. Data presented as a comparison between the increase in the lysosomal abundance of GPDs upon CLN3 loss in LAMP1-3×HA tagged lysosomes relative to that in TMEM192-3×HA tagged lysosomes (mean ± s.e.m.; $n = 4$ biologically independent samples). n.s: non-significant; Two-tailed unpaired t-test.

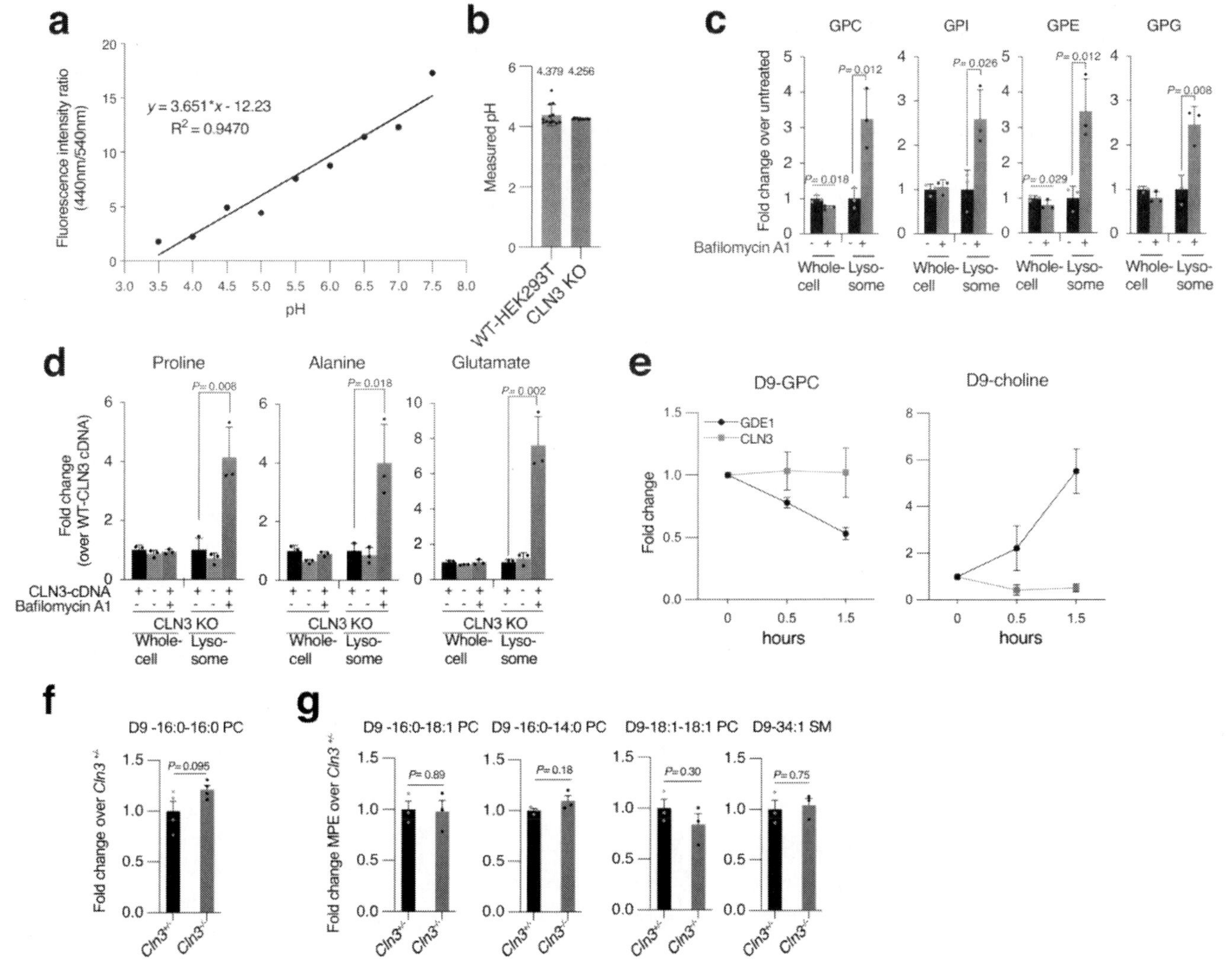

Extended Data Fig. 4 | Testing the effects of CLN3 loss on lysosomal pH and cellular lipid metabolism. a to **d**, CLN3 loss does not increase lysosomal pH. **a**, A standard calibration curve of the ratiometric dye LysoSensor Yellow/Blue DND-160. See methods for experimental details. **b**, Lysosomal pH in wild-type and CLN3 KO HEK293T cells as calculated using the standard curve in **a**. Data are presented as mean ± s.d., *n = 12 biologically independent samples*. **c**, Targeted analyses of the fold changes in whole-cell and lysosomal levels of GPDs upon treatment with 500 nM Bafilomycin A1 for 6 h in CLN3 expressing HEK293T cells. Data are presented as mean ± s.d., *n = 3 biologically independent samples*, (Two-tailed unpaired *t*-test). **d**, The levels of several amino acids whose egress from lysosomes is sensitive to the proton gradient across the lysosomal membrane are not affected by CLN3 loss. The levels of proline, alanine, and glutamate in whole cells and lysosomes were compared between CLN3 KO cells with and without CLN3 cDNA addback and upon treatment with 500 nM Bafilomycin A1 (BafA1) for 6 h. Data are presented as mean ± s.d., *n = 3 biologically independent samples*, (Two-tailed unpaired *t*-test). **e**, The recombinant CLN3 protein does not have glycerophosphodiesterase activity. Deuterated GPC (D9-GPC) was incubated with recombinant 3×Flag-CLN3 or the positive control glycerophosphodiesterase1 (3×Flag-GDE-1) for the indicated time points. The extent of D9-GPC hydrolysis was determined by measuring the decline in its level and the increase in the levels of the product D9-choline. Data presented as mean ± s.e.m. of *n* = 3 biologically independent samples. **f**, CLN3 loss does not decrease tracer uptake. Fold change in tracer levels (D9-16:0-16:0 PC) normalized to total protein for samples measured in Fig. 4f and g. Data presented as mean ± s.e.m. of *n* = 4 biologically independent samples. **g**, CLN3 loss does not affect the biosynthesis or turnover of PC and sphingomyelin (SM) in cells. Free D9-choline was used as a tracer. Data are presented as fold changes in the whole-cell molar percent enrichment (MPE) of D9-choline-containing lipids in cortical neuron cultures prepared from *Cln3*$^{-/-}$ mice relative to those from *Cln3*$^{+/-}$ animals (mean ± s.e.m., *n* = 3 biologically independent samples, (Two-tailed unpaired *t*-test)).

nature portfolio

Corresponding author(s): Monther Abu-Remaileh

Last updated by author(s): 06/24/2022

Reporting Summary

Nature Portfolio wishes to improve the reproducibility of the work that we publish. This form provides structure for consistency and transparency in reporting. For further information on Nature Portfolio policies, see our Editorial Policies and the Editorial Policy Checklist.

Statistics

For all statistical analyses, confirm that the following items are present in the figure legend, table legend, main text, or Methods section.

n/a	Confirmed	
☐	☒	The exact sample size (*n*) for each experimental group/condition, given as a discrete number and unit of measurement
☐	☒	A statement on whether measurements were taken from distinct samples or whether the same sample was measured repeatedly
☐	☒	The statistical test(s) used AND whether they are one- or two-sided *Only common tests should be described solely by name; describe more complex techniques in the Methods section.*
☒	☐	A description of all covariates tested
☒	☐	A description of any assumptions or corrections, such as tests of normality and adjustment for multiple comparisons
☐	☒	A full description of the statistical parameters including central tendency (e.g. means) or other basic estimates (e.g. regression coefficient) AND variation (e.g. standard deviation) or associated estimates of uncertainty (e.g. confidence intervals)
☐	☒	For null hypothesis testing, the test statistic (e.g. *F*, *t*, *r*) with confidence intervals, effect sizes, degrees of freedom and *P* value noted *Give P values as exact values whenever suitable.*
☒	☐	For Bayesian analysis, information on the choice of priors and Markov chain Monte Carlo settings
☒	☐	For hierarchical and complex designs, identification of the appropriate level for tests and full reporting of outcomes
☒	☐	Estimates of effect sizes (e.g. Cohen's *d*, Pearson's *r*), indicating how they were calculated

Our web collection on statistics for biologists contains articles on many of the points above.

Software and code

Policy information about availability of computer code

Data collection	All softwares for data collection are publicly or commercially available: Tune version 2.9, Xcalibur 4.1, MetaMorph 7, ZEN 3.2
Data analysis	All softwares for data analysis are publicly or commercially available: Microsoft Excel v16.61 and 16.62, GraphPad Prism v9.0, XCalibur v4.1, ImageJ v1.52, CompoundDiscoverer v3.1 , Spectronaut Professional v13.10, LipidSearch v4.2.21, TraceFinder software v5.0, R version 3.4.1 and RStudio server version 1.1.463,

For manuscripts utilizing custom algorithms or software that are central to the research but not yet described in published literature, software must be made available to editors and reviewers. We strongly encourage code deposition in a community repository (e.g. GitHub). See the Nature Portfolio guidelines for submitting code & software for further information.

Data

Policy information about availability of data

All manuscripts must include a data availability statement. This statement should provide the following information, where applicable:

- Accession codes, unique identifiers, or web links for publicly available datasets
- A description of any restrictions on data availability
- For clinical datasets or third party data, please ensure that the statement adheres to our policy

The mass spectrometry proteomics data were deposited to the ProteomeXchange Consortium via the PRIDE partner repository with the dataset identifier PXD018624. Known lysosomal proteins in our proteomics analysis were defined based on Gene Ontology Cellular Component or UniProt sub-cellular localization annotation (https://www.uniprot.org/). The conditional LysoTag mouse was deposited at the Jackson Laboratory (Strain #035401). Other unique biological materials in the form of plasmids or cell lines are available from corresponding author upon request. Detailed lipidomics and metabolomics data are provided in

Supplementary Tables 3 and 4, respectively. Other data generated are available from the corresponding author upon request. Source data are provided with this paper.

Field-specific reporting

Please select the one below that is the best fit for your research. If you are not sure, read the appropriate sections before making your selection.

☒ Life sciences ☐ Behavioural & social sciences ☐ Ecological, evolutionary & environmental sciences

For a reference copy of the document with all sections, see nature.com/documents/nr-reporting-summary-flat.pdf

Life sciences study design

All studies must disclose on these points even when the disclosure is negative.

Sample size	No formal sample-size calculations were performed. However, based on previously published work and literature in the field that have conducted similar experiments as well as the number of replicates sufficient for statistical analyses, we used a minimal sample size of n = 3 for quantitative experiments. For some experiments, where we expected higher variance, generally due to lower amounts of starting material (e.g. LysoIP), we included more samples per condition. Taking this into account for in vivo experiments, the final exact numbers were determined based on the availability of each genotype, experimental design and animal housing conditions. We observed that sample sizes of 3 to 6 animals per group was adequate sample size and sufficient to obtain statistical power. For cell based experiments, each independent culture of cells was defined as biological replicate.
Data exclusions	In rare cases where we had very poor detection across the majority of metabolites the whole sample was excluded.
Replication	For most of our figures, at least three independent experiments were performed and have not encountered discrepancy between results and conclusions. Extended data figure 1A was reproduced only two times. Extended data figures. 1E and 1F, the transmission electron microscopy (TEM) provide the same observations from two animals per group.
Randomization	To ensure rapid processing, LysoIP protocols were performed one sample at a time and usually by two people and samples were randomized across conditions. To prevent any systematic bias from possible column carryover, sample order was randomized in LC/MS runs. For pairwise comparisons, samples were processed in internally controlled pairs (i.e. Wild-type vs. KO), while matching sex and age. For other experiments where samples were not allocated randomly, group allocation and randomization were unnecessary because all samples were measured independently in the same way in an internally controlled manner.
Blinding	Sample blinding was not performed for most experiments as they did not involve formal qualitative scoring. Furthermore, all measurement modalities were quantitative and all samples and experiments were performed in the same way regardless of group or treatment. For in vivo metabolomic and proteomic experiments as well as for analysis of patient CSF samples, the investigators were blinded during data acquisition since mouse or patient number was used as identifier, respectively. Immunofluorescence (IF) assessments were qualitative and based on identifiable features. In at least one replicate for each IF experiment, samples were independently evaluated by at least one other investigator whose interpretation supported the conclusions in the manuscript.

Reporting for specific materials, systems and methods

We require information from authors about some types of materials, experimental systems and methods used in many studies. Here, indicate whether each material, system or method listed is relevant to your study. If you are not sure if a list item applies to your research, read the appropriate section before selecting a response.

Materials & experimental systems

n/a	Involved in the study
☐	☒ Antibodies
☐	☒ Eukaryotic cell lines
☒	☐ Palaeontology and archaeology
☐	☒ Animals and other organisms
☐	☒ Human research participants
☐	☒ Clinical data
☒	☐ Dual use research of concern

Methods

n/a	Involved in the study
☒	☐ ChIP-seq
☒	☐ Flow cytometry
☒	☐ MRI-based neuroimaging

Antibodies

Antibodies used	Antibody Target:Company Catalogue number/clone number Dilution used in this study LAMP2: Santa Cruz Biotechnology: sc-18822/ H4B4 1:1000 LAMP1: Developmental Studies Hybridoma Bank: 1D4B 1:1000 VDAC1: Cell Signaling Technology: CST-4661/ D73D12 1:2000

GOLGIN-97: Cell Signaling Technology: CST-13192/ D8P2K 1:500
CALR: Cell Signaling Technology: CST-12238/ D3E6 1:2000
CTSB: Cell Signaling Technology: CST-31718/ D1C7Y 1:1000
CTSC: Santa Cruz Biotechnology: sc-74590/ D-6 1:1000
HA epitope: Cell Signaling Technology: CST-2367/ 6E2 1:2000
FLAG epitope: Cell Signaling Technology: CST-14793/ D6W5B 1:2000
Catalase: Cell Signaling Technology: CST-14097/ D5N7V 1:1000
CTSD: R&D Systems: AF1029/ polyclonal 1:1000
NeuN: Cell Signaling Technology: CST-24307/ D4G4O 1:1000
AKT: Cell Signaling Technology: CST-4685/ 11E7 1:3000
Vinculin: Santa Cruz Biotechnology: sc-73614/ 7F9 1:3000
p70 S6 Kinase: Cell Signaling Technology: CST-9202/ polyclonal 1:1000
Histone H3: Cell Signaling Technology: CST-4499/ D1H2 1:500

Secondary antibodies for IF: Goat anti-Rat Alexa Fluor 488 (Thermo A-11006), Goat anti-rabbit Alexa Fluor 488 (Thermo A-21206), Goat anti-Mouse Alexa Fluor 647 (ThermoA-21240), Goat anti-Mouse Alexa Fluor 488 (Thermo, A-11001), and Goat anti-rabbit Alexa Fluor 594 (Thermo, A32740).
Secondary antibodies for immunoblotting: Anti-rabbit HRP (CST, 7074), anti-mouse HRP (CST, 7076), anti-rat HRP (CST, 7077)

Validation

All antibodies used in this study are acquired from commercial sources and were validated for specificity and species reactivity by the manufacturer. The information are readily available on the website of the manufacturer listed here:
Antibody Target Direct link to manufacturer's website

LAMP2 https://www.scbt.com/p/lamp-2-antibody-h4b4

LAMP1 https://dshb.biology.uiowa.edu/1D4B

VDAC1 https://www.cellsignal.com/products/primary-antibodies/vdac-d73d12-rabbit-mab/4661

GOLGIN-97 https://www.cellsignal.com/products/primary-antibodies/golgin-97-d8p2k-rabbit-mab/13192?site-search-type=Products&N=4294956287&Ntt=golgin-97&fromPage=plp&_requestid=789344

CALR https://www.cellsignal.com/products/primary-antibodies/calreticulin-d3e6-xp-rabbit-mab/12238

CTSB https://www.cellsignal.com/products/primary-antibodies/cathepsin-b-d1c7y-xp-rabbit-mab/31718

CTSC https://www.scbt.com/p/cathepsin-c-antibody-d-6

HA epitope https://www.cellsignal.com/products/primary-antibodies/ha-tag-6e2-mouse-mab/2367?site-search-type=Products&N=4294956287&Ntt=ha+epitope+2367&fromPage=plp&_requestid=789757

FLAG epitope https://www.cellsignal.com/products/primary-antibodies/dykddddk-tag-d6w5b-rabbit-mab-binds-to-same-epitope-as-sigma-s-anti-flag-m2-antibody/14793?site-search-type=Products&N=4294956287&Ntt=flag+epitope+14793&fromPage=plp&_requestid=789909

Catalase https://www.cellsignal.com/products/primary-antibodies/catalase-d5n7v-rabbit-mab/14097?site-search-type=Products&N=4294956287&Ntt=catalase+14097&fromPage=plp&_requestid=790017

CTSD https://www.rndsystems.com/products/mouse-cathepsin-d-antibody_af1029

NeuN https://www.cellsignal.com/products/primary-antibodies/neun-d4g4o-xp-rabbit-mab/24307?site-search-type=Products&N=4294956287&Ntt=neun+24307&fromPage=plp&_requestid=790222

AKT
https://www.cellsignal.com/products/primary-antibodies/akt-pan-11e7-rabbit-mab/4685?site-search-type=Products&N=4294956287&Ntt=akt+4685%5C&fromPage=plp&_requestid=790281

Vinculin https://www.scbt.com/p/vinculin-antibody-7f9

p70 S6 Kinase https://www.cellsignal.com/products/primary-antibodies/p70-s6-kinase-antibody/9202?site-search-type=Products&N=4294956287&Ntt=p70+s6+kinase+9202&fromPage=plp&_requestid=790500

Histone H3 https://www.cellsignal.com/products/primary-antibodies/histone-h3-d1h2-xp-rabbit-mab/4499?site-search-type=Products&N=4294956287&Ntt=histone+h3+4499&fromPage=plp&_requestid=790630

Lamp1 1D4B was confirmed via IF to overlap with over-expressed TMEM192, a marker of lysosomes.

Eukaryotic cell lines

Policy information about cell lines

Cell line source(s)

HEK-293T cells were from ATCC

Authentication

STR profiling

Mycoplasma contamination	Test routinely and negative results obtained
Commonly misidentified lines (See ICLAC register)	No commonly misidentified cell lines were used in the study

Animals and other organisms

Policy information about studies involving animals; ARRIVE guidelines recommended for reporting animal research

Laboratory animals	Experiments were performed on the following mouse strains: LysoTag mice of the C57BL/6J background (original chimeras were backcrossed for at least three generations). Cln3-/- mice on C57BL/6J background. CMV-Cre mice on C57BL/6J background. Age and sex are all indicated in the figure legend for each experiment. Male and female mice were used with an average age of 5.5 weeks to 7 months and age- and sex-matched in each experiment. All mice were maintained on a standard light-dark cycle with access to food and water ad libitum. Animals were fasted for the fasting experiment in Fig. 1. The animal room had a controlled temperature of 22°C, and humidity (around 50%). Cages were cleaned every 4-5 days and supplies of water and food were checked daily.
Wild animals	none
Field-collected samples	none
Ethics oversight	All procedures involving mice were carried out in accordance with the approved guidelines by the Institutional Animal Care and Use Committee at MIT, the Whitehead Institute for Biomedical Research and Stanford University.

Note that full information on the approval of the study protocol must also be provided in the manuscript.

Human research participants

Policy information about studies involving human research participants

Population characteristics	See reference Dang Do, A. N. et al. (2021) Genet Med 23, 751-757. Characteristics of study participants are also included in Supplementary Table 5 of this manuscript.
Recruitment	Recruitment of study participants were approved by the institutional IRB and included self-referrals, referrals from family support organizations, providers, and through clinicaltrials.gov. The studies enroll individuals of any age, sex, race, ethnicity, and disease severity who meet the eligibility criteria defined in the respective website at clinicaltrials.gov and who are able to travel to NIH Clinical Center without increased health risk. Self-selection ascertainment bias is an inherent part of any clinical trial (the participant makes a choice to participate), however, we don't expect this to cause a systemic bias and impact our study.
Ethics oversight	The Eunice Kennedy Shriver National Institute of Child Health and Human Development, National Institutes of Health IRB approved the natural history studies. Individuals of any age, sex, and race who have at least one variant in CLN3 and clinical symptoms suggestive of CLN3 disease were enrolled following appropriate provision of consent and assent. For minors, parents or legally authorized representatives provided permission and consent.

Note that full information on the approval of the study protocol must also be provided in the manuscript.

Clinical data

Policy information about clinical studies

All manuscripts should comply with the ICMJE guidelines for publication of clinical research and a completed CONSORT checklist must be included with all submissions.

Clinical trial registration	NCT033007304, NCT02931682, NCT00001721, NCT00344331
Study protocol	The samples were collected as parts of observational, natural history studies. The protocols are available from authors upon reasonable requests.
Data collection	The samples were collected at the NIH Clinical Center as parts of natural history studies. In particular, the CLN3 samples were collected from 2017-2020.
Outcomes	The samples were collected as parts of natural history studies that included extensive phenotyping and biomarker discoveries.

Untargeted metabolomic and lipidomic profiling for Golgi molecular content via immunoprecipitation (Golgi-IP)

Wentao Dong, Rotimi Fasimoye, Raja Nirujogi, Eshaan Rawat, Miharu Iguchi, Uche Medoh, Kwamina Nyame, Dario Alessi, Monther Abu-Remaileh

Organelle-specific metabolic dysfunctions have been implicated in neurodegenerative diseases, but it is challenging to achieve high resolution for compartmentalized metabolism and detect metabolites at low nanomolar concentrations in small organelles such as lysosomes and Golgi. To quantitatively assess novel subcellular proteins and compartmentalized metabolism, we developed a high-resolution mass spectrometry (HRMS) method in conjunction with a technique to harvest organelle-specific metabolites and lipids via immunoprecipitation (IP). In addition to uncovering novel Golgi proteins, we applied our HRMS-Golgi IP workflow to study the Golgi-specific metabolome and identified a class of uridine-diphosphate (UDP) sugars and their derivatives that were specifically enriched in the Golgi. Moreover, we validated Golgi-specific reduction of a UDP species after genetically knocking out its transporter on the Golgi. Further isotope tracing analysis using ^{13}C-labeled glucose suggests a potentially new metabolic coupling between UDP species within the Golgi. In addition, we also performed untargeted lipidomics using Golgi IP and identified a group of phospholipids to be selectively enriched in the Golgi. Our work established a robust HRMS-IP method to investigate subcellular metabolism and provided novel insights into the Golgi proteome, metabolome and lipidome. These insights can also lead to discoveries of new metabolic reactions within cellular compartments that remain elusive by conventional whole-cell approaches.

AUTHOR INDEX

AIChE
120 Wall Street, FL 23
New York, NY 10005-4020

ISBN 978-1-7138-7920-6

9 781713 879206

CURRAN ASSOCIATES INC.
proceedings.com